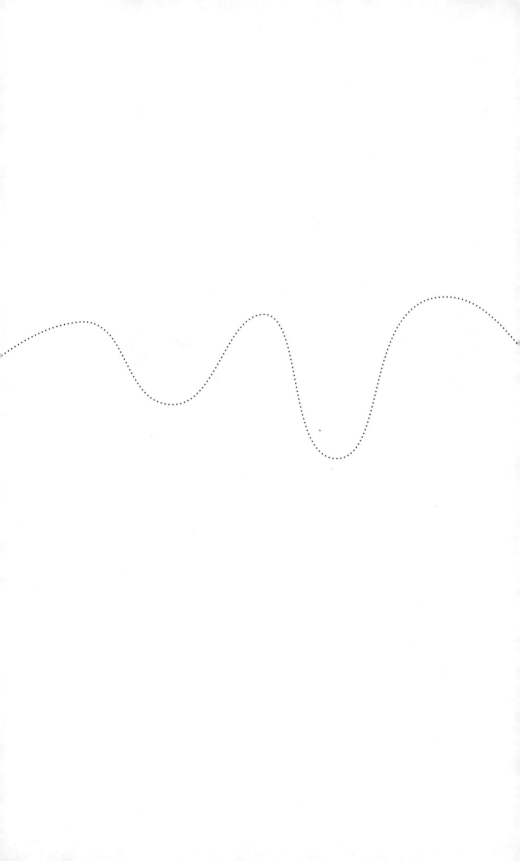

PLANET OF THE

BUGS

EVOLUTION AND THE RISE OF INSECTS

* * * * * * * * **Scott Richard Shaw** * * * * * * * *

THE UNIVERSITY OF CHICAGO PRESS   CHICAGO AND LONDON

Scott Richard Shaw is professor of entomology and Insect Museum curator at the University of Wyoming, Laramie.

The University of Chicago Press, Chicago 60637
The University of Chicago Press, Ltd., London
© 2014 by The University of Chicago
All rights reserved. Published 2014.
Printed in the United States of America

23 22 21 20 19 18 17 16 15 14   2 3 4 5

ISBN-13: 978-0-226-16361-1 (cloth)
ISBN-13: 978-0-226-16375-8 (e-book)
DOI: 10.7208/chicago/9780226163758.001.0001

Library of Congress Cataloging-in-Publication Data
Shaw, Scott R. (Scott Richard), author.
Planet of the bugs : evolution and the rise of insects /
Scott Richard Shaw.
pages cm
Includes bibliographical references and index.
ISBN 978-0-226-16361-1 (cloth : alk. paper) —
ISBN 978-0-226-16375-8 (e-book)
1. Insects—Evolution. I. Title.
QL468.7.S53 2014
595.7—dc23

                                        2013050775

♾ This paper meets the requirements of
ANSI/NISO Z39.48-1992 (Permanence of Paper).

This book is dedicated to my wife, Marilyn, for her patience, support, and most especially understanding. Thanks for letting me keep bugs in the freezer. I couldn't have done it without you.

Contents

A long-horned beetle (family Cerambycidae) perches on a melastome leaf after a recent rain at San Ramon forest in Costa Rica.

Prologue: Time Travel with Insects

Time flies like an arrow. Fruit flies like bananas.

GROUCHO MARX

Late one October afternoon I was walking along a trail through the rain forest at San Ramon Biological Reserve in Costa Rica, pondering the nature of time and wishing for a time machine. The earth's tropical wet forests do not display obvious seasonality so you would never guess the day, month, or year by looking at the surrounding mossy vegetation, and they emanate such an ancient and timeless green aura that it's easy to imagine traveling back in time, thousands, hundreds of thousands, or even millions of years. The scent of decaying vegetation and fungus permeated the wet air, and the forest was both figuratively and literally crawling with insects. Around my feet, abundant small flies and other insects enjoyed the bounty of rotting fruits fallen to the forest floor. I recalled Groucho's classic punch line in *Duck Soup*: "Fruit flies like bananas."

As the songs of frogs, katydids, crickets, and cicadas emanated from the forest, my boots sloshed along the pathway. Typical of San Ramon, it had been raining all day, the trail oozed treacherously slick with slippery mud, and water was everywhere. On mushroom caps sprouting from a rotting log by the trail, silvery droplets rolled to the edge, clung briefly shimmering—then fell away. The sounds of water were all around, bubbling and gurgling over mossy rocks in the river, chattering in nameless streams and rivulets. A light mist was still falling, and the emerald vegetation, dappled in a hundred shades of green, was dripping and glistening with raindrops. The trees at San Ramon were matted with mosses, lichens, and ferns that had absorbed water, sponge-like, later releasing it gradually long after the rain had stopped. Up in the forest canopy, bromeliads had collected rainwater in their concave leaf bases, forming miniature ponds for numerous tree frogs, salamanders, and hundreds of kinds of aquatic insects. Water, everywhere water, was dripping.

Although I took my time walking through the slick mud, placing each step with care, maybe I should have been daydreaming less. The San Ramon forest contains thousands of insect species, many of them still new and undiscovered, but unfortunately, it also contains lots of venomous snakes and some deadly ones as well—and like Indiana Jones, I hate snakes. Even so, I was enjoying myself immensely, finding fascinating plants and insects all around. Then unexpectedly, just past a bend, down a gentle slope near a small gurgling stream, I came across a small melastome tree. There, on a large leaf about four feet off the ground, I found my time machine. It was about three inches long and deep mahogany brown, with long curved antennae adorned with dewdrops from the recent rain. Standing motionless on the leaf, it rested securely, supported by its six multijointed legs that formed two perfect tripods. My time machine was a long-horned beetle.

As the rain began to fall more heavily at San Ramon, new beads of water started dripping from the beetle. Its tough armored exterior would not allow them to penetrate. The beetle's body plan is extremely different from our own, with its rigid skeleton on the outside rather than the inside, and this external skeleton, as well as the beetle's multijointed legs, send us a message from the shallow oceans of the Cambrian period, roughly 541 million years ago. Atmospheric oxygen levels increased at that time, animal metabolism accelerated, and marine predators became faster and nastier. In response, the common ancestor of all living arthropods (insects and their nearest relatives) evolved external skeletons, which provide protection from the environment, defense against predators, and sites of muscle attachment that increase mobility. They evolved jointed legs as well, which are also exceptionally useful for defending against predators, as well as for food-gathering, mating, and dispersal. Both features have been quite fashionable ever since.

The beetle remained motionless on the melastome leaf, but on close inspection I noticed that its abdominal body segments were contracting then expanding slightly. The beetle was breathing. Along the sides of its body, minute holes (spiracles) allowed molecules of air to rush in. If we could shrink ourselves down to the size of a few microns, we might follow those oxygen molecules on their travels through the beetle's spiracles, through a series of large tubes (tracheae), and finally through a series of smaller tubes (tracheoles) that branch out in all directions, and get smaller and smaller until they approach all the living cells in the insect's body. The fact that the beetle breathes air tells us that in the Silurian period, roughly 444 million years ago, ancestral arthropods first colonized the land and developed air-breathing tracheal respiration systems that have been passed onwards to all modern insects, even those that live in fresh water and breathe through gills. Their innovative breathing apparatus allowed Silurian arthropods to exit the oceans to avoid marine predators, forage for food along the shorelines, and use those unoccupied beaches as a safe place to mate and lay eggs. It also prepositioned the arthropods to be the first animal group to successfully diversify among the earliest Lilliputian plant communities, which developed during the Late Silurian.

I examined the beetle's form more closely. Its body was divided into three functional regions: a head up front, from which its hornlike antennae swept to the sides, and small multifaceted eyes gazed back at me; a central thorax, to which its legs were attached; and lastly, a multisegmented abdomen. These are the features that define an insect. They evolved sometime in the Late Silurian or Early Devonian, about 419 million years ago. During this time, plant communities became taller and more diverse, the first insects evolved among the mossy soils at the bases of the first archaic trees, and six-leggedness established itself as a versatile and stable means of walking, standing, and running on this planet, allowing the insects to become highly successful scavengers. By the Late Devonian period, 360 million years ago, springtails, jumping bristletails, and silverfish were abundant among the accumulating litter of decaying plant materials. The feeding activities of these scavengers created and conditioned organic soils, which enabled the earth's forests to evolve and expand into the interior from the shorelines. With tiny but triumphant steps, microscopic insects marched inland as the plant communities extended their colonization of the continents.

Along the backside of the beetle ran a hairline seam, which indicated the presence of wings: a message from the common ancestor of all flying insects, which first evolved in the Carboniferous period, about 354 million years ago. When the insects invented wings there were no other flying animals—no birds, bats, pterodactyls, or gliding squirrels—and they completely mastered the air for more than 150 million years before any other organisms evolved the ability to fly or could chase them in the air. The advantage of flight was certainly one of the foremost factors contributing to the explosive proliferation of insect species on this planet. But wings not only allowed insects to disperse and colonize distant new habitats, they also played important roles in courtship and mating, predator avoidance, food acquisition, and thermoregulation. By the late Carboniferous Period, tall forests had spread across the Earth's continents. They were a glittering fairyland of curious flying insects: banded, spotted, and net-winged paleodicytopterans; dragonfly-like griffenflies and protodonatans; ancestral mayflies; and even sundry forms of flying roaches that fluttered and glided among the treetops.

Preparing to fly, the beetle extended its rigid, shell-like front wings outward and unfolded its membranous hind wings. This style of wings is, in fact, the beetles' key innovation, evolved some 260 million years ago in the Permian period. Only the hind wings power beetle flight; the modified armored front wings allow the delicate hind wings to be put away, hidden and protected, when not in use. These hard front wings, unique to beetles, also protect the beetles when they're not flying and give them a more streamlined body profile, which enables them to crawl into cracks and crevices, under loose bark, in leaf litter, and among woody debris. While the beetles seem to fly clumsily with the unfolded hind wings, they are efficient enough to allow the beetles to disperse widely in search of mates and food, and to colonize new habitats.

Some of those ancient Permian beetle ancestors also evolved a very useful feeding behavior, retained even today by the modern longhorned wood-boring beetle. Their immature forms, larvae, developed the habit of boring deep into the woody trunks of dead trees to feed there on the fungal growth and decaying plant tissues. This sheltered them from increasingly adept warm-blooded predators throughout the Permian, and on through the age of the dinosaurs, the Mesozoic era from the Triassic through the Cretaceous periods. It may have also served to buffer them from environmental change. At the end of the Permian, about 252 million years ago, the ancient continents collided to form the supercontinent of Pangaea. Coastlines and marine habitats were severely disrupted, perhaps triggering a mass extinction of species greater than any other extinction event so far. Terrestrial habitats became hotter and drier than before, but this only seemed to stimulate the beetles' success. While there were only 5 families of primitive beetle-like insects in the Late Permian, by the Late Triassic (around 220 million years ago) at least 20 families consisting of 250 species of true beetles had evolved. Over the course of the following Jurassic period, despite living among hungry dinosaurs, beetle numbers continued to skyrocket; at least 600 species in 35 families have been identified in middle-Mesozoic fossils.

Back along the trail at San Ramon, the beetle flexed its wings again and took a short, buzzing flight to a nearby yellow flower. After a few moments, it slowly began to chew, lazily indulging in a high-protein pollen meal. The sight of insects feeding on flowers is so commonplace in our modern world that we tend to forget how unusual it really is, geologically speaking. The ancient Permian proto-beetles didn't visit flowers because they didn't exist yet, at least not in the sense that we understand them. The flowering plants that currently dominate the landscape, known by botanists as angiosperms, did not evolve until the Early Cretaceous period, around 126 million years ago. Perhaps sometime during that period an ancient beetle first visited a flower, maybe in the shadow of a *Tyrannosaurus* or *Triceratops* dinosaur, and found the pollen tasty. Over the Late Cretaceous, this time in tandem with the flowering plants' early diversification, the richness of beetle species skyrocketed again, particularly among many of the plant-feeding beetle groups that survived the period and now dominate our modern world: leaf beetles, weevils, scarabs, click beetles, metallic wood-boring beetles, and, notably, the long-horned beetles, the family into which our time-machine beetle is classified.

Maybe, 66 million years ago, some Cretaceous beetles were busily feeding on pollen from ancient magnolia blossoms when a sound from above caused a nearby *Tyrannosaurus* to glance briefly skyward and see a massive asteroid hurtling toward the earth—a catastrophe which brought the time of the giant dinosaurs to a close and marked the end of the Cretaceous. Global winter ruled for a time, killing off not only dinosaurs but also perhaps many kinds of small marine organisms. But deep in rotting tree trunks and elsewhere, the larvae of many beetles survived, completed their metamorphoses, and emerged into a brave new world without giant dinosaurs. Over ensuing millions of years, some of those survivors lived on, evolved, and diversified to become the most species-rich animal group in the world today.

Our beetle flew into the variegated green of the primeval forest, perhaps to seek others of its kind, and in doing so, to replicate its many messages from the past. Its gentle buzzing was lost among the sounds of dripping water and the increasing rainfall. This beetle is gone now, but many others remain. We can only estimate their numbers. Studies by Smithsonian entomologist Terry Erwin indicate that there are millions, perhaps tens of millions, of different kinds of beetles in our tropical forests, most of them still unnamed. And that's just one major insect order. Many other kinds of insects are hyperdiverse, such as the moths, butterflies, true flies, wasps, and true bugs. This hyperdiversity is the rich historical legacy of the Cenozoic era. Over the past 66 million years, as the insects have continued to diversify along with the flowering plants, our tropical rain forests have evolved into the most biologically complex and diverse ecosystems ever to arise. Why are there so many different kinds of insects, and why do they dominate terrestrial ecosystems? Science has unlocked an extraordinary number of mysteries, and the story of the insects' rise can now be read over hundreds of millions of years of earth's history. The messages are written there in the rocks, the forests, and the insects, for those who choose to read them.

1 The Buggy Planet

It is for me a stunning fact that while the physical surface of the earth has been thoroughly explored, so that virtually every hilltop, tributary, and submarine mount has been mapped and named, the living world remains largely unknown. As few as ten percent of the species of insects and other invertebrate animals have been discovered and given scientific names.

EDWARD O. WILSON, *The High Frontier*

All things have a root and a top,
All events an end and a beginning;
Whoever understands correctly
What comes first and what follows
Draws nearer to Tao

BARRY HUGHART, *Bridge of Birds*

Earth is a very buggy planet. Nearly one million distinct living species, different kinds of insects, have been discovered and named so far. From A to Z, they overwhelm us with their diversity: ants, birdwing butterflies, cockroaches, dung beetles, earwigs, flies, grasshoppers, head lice, inchworms, June beetles, katydids, ladybugs, mantises, net-winged midges, owlflies, periodical cicadas, queen termites, royal palm bugs, sawflies, thrips, underwing moths, velvety shore bugs, webspinners, xyelid sawflies, ypsistocerine wasps, and zorapterans. But that is just the tip of the iceberg, the door to the hive. Most of the insect species haven't even been given a name, and scientists estimate that the number of different kinds of insects living in tropical forests is perhaps in the tens of millions.[1] Whether you adore them or abhor them, their diversity and ecological success is impressive.

Insects are so successful that it's not much of an exaggeration to say that they literally rule the planet. Our egos allow us to think that we humans rule earth, with our cities, our technology, and our civilizations, but we seem to be doing more to destroy the planet than to improve it, and we are like one superabundant pest species run amok over the globe. If humans were to suddenly become extinct, the living conditions for most species would be greatly improved with only a few exceptions, such as human body lice and crab lice. On the other hand, if all the insects became extinct, in the words of Edward O. Wilson, the famous Harvard entomologist, "the terrestrial environment would collapse into chaos."[2] Human civilizations have only recently developed over the last several thousand years. Insects have successfully coevolved with terrestrial ecosystems over the last four hundred million years. They are ecologically essential as scavengers, nutrient recyclers, and soil producers, feeding on and utilizing virtually every kind of imaginable organic material. Six-legged detritivores consume dead plants, dead animals, and animal droppings, greatly increasing the rates at which these materials biodegrade. Insects, as both predators and parasitoids, are keystone organisms that feed upon and reduce populations of other kinds of plant-feeding and scavenging insects. They are also their own worst enemies: most kinds of insects have populations that are kept in check by the feeding activities of other insects.

Over the past 120 million years, insects have coevolved and explosively diversified in tandem with the angiosperms—the dominant forms of plant diversity in modern ecosystems. They are essential as pollinators and seed-dispersers for most of the flowering plants, whose communities would be vastly diminished if all plant-associated insects were eliminated. We often tend to think of plant-feeding insects in general as pests, but I like to point out that only a miniscule small fraction (less than 1 percent) of the total number of insect species are actually significant pests. In fact, most of the plant-feeding insects should be considered beneficial for two reasons. First, they reduce the reproductive output of particular plants by putting stress on them. That sounds bad if the plant is an agricultural crop, but in a natural setting, such as a tropical forest or a mountain meadow, that plant feeding has a very desirable outcome. It prevents particular plant species from becoming superabundant and weedy, allowing vastly more species to coexist in much smaller spaces. Plant-feeding insects are a driving force in the evolution of plant community species richness, and so the extraordinary plant diversity of tropical habitats is largely due to insect diversity, not despite it. Second, but of no less importance, the majority of plant-feeding insects are themselves edible to other kinds of wildlife. Many insects are a fundamental and nutritious food source for most kinds of vertebrate species, including fish, amphibians, reptiles, birds, and most mammals, including primates and even humans. Not many organisms totally depend on humans for their continued existence, but a large part of living plants and terrestrial animals depend partly or entirely on insects for their survival.

Whether or not they rule the planet, insects certainly have largely overrun it. They can be found in abundance in virtually every kind of terrestrial habitat, from tropical rain forests to deserts, in meadows and prairies, from sea shorelines to alpine tundra and Andean páramo. Aquatic insects not only inhabit mountain streams, rivers, waterfalls, seepages, lakes, ponds, swamps, and salt marshes, but they even occupy mud puddles, sewage ponds, craters in rocks, tree holes, pitcher plant leaves, and bromeliad leaf bases more than a hundred feet above the forest floor. Semiaquatic insects exploit the force of surface tension to skate across still ponds and lakes, while the ocean water strider, genus *Halobates*, has been seen walking on the ocean surface hundreds of miles at sea. Clouds of millions of African migratory locusts have flown across the entire Atlantic Ocean to land in the Caribbean Islands. The insect macro-societies, ants and termites, are essential soil movers in the Amazon basin, where their biomass outweighs the biomass of vertebrates. But sheer insect abundance is not strictly a tropical phenomenon. Even near the Arctic Circle, the combined weight of biting flies and midges outweighs that of the mammals.

Insects and their relatives have evolved and adapted to some of the most extreme conditions on the planet. Stoneflies have been recorded at an elevation of 5,600 meters in the Himalayas, while subterranean species of beetles, crickets, and cockroaches have adapted to life in caves deep underground. Some aquatic stream beetles breathe across the surface of an air bubble and can stay underwater indefinitely. Brine flies, shore flies, seaweed flies, and deer flies have developed extreme tolerance for high levels of salt and live in salt marshes and salt flats and along ocean shorelines. Springtails have evolved antifreeze compounds in their blood, and some are among the most abundant organisms on sub-Antarctic islands. At high elevations worldwide, species of icebugs, springtails, snow scorpionflies, and some flightless

FIGURE 1.1. Common denizens of the leaf litter, springtails (order Collembola) tolerate many environmental extremes. (Photo by Kenji Nishida.)

tipulid flies are active on the frozen surfaces of snow fields and glacial ice. Living chironomid midge larvae have been dredged up from the depths of Lake Baikal in Russia, where they have adapted to a low-oxygen environment by evolving hemoglobin-like blood pigments. The adaptability of water boatmen bugs is remarkable: some inhabit salty water below sea level in Death Valley, California, while others live high in the Himalayan Mountains. Some swim in frigid water under ice, while others thrive in hot springs at temperatures up to 35°C. The Yellowstone hot springs alkali fly develops in the edges of scalding hot water pools with temperatures up to 50°C. Other fly larvae living in arctic ponds are known to survive winter cold temperatures as low as –30°C. One of the most impressive organisms is the South African chironomid midge fly, *Polypedilum vanderplanki*, which has adapted to extreme drought conditions by evolving cryptobiosis—a suspended-animation condition where larvae become dehydrated and tolerant to the most extreme conditions. It has been reported that these dehydrated fly larvae can tolerate immersion in boiling water as well as being dipped into liquid helium.

Most insect species are not nearly so tolerant of a wide range of extremes, and indeed, many fresh water stream insects have such a narrow range of acceptable conditions of water temperature and oxygen levels that they are very valuable to us as bioindicators of good water quality. On the other hand, hundreds of thousands of tropical plant-feeding insects have evolved physiologies that allow them to feed on and metabolize plants that are highly toxic to mammals and most other animals. Many tropical caterpillars are able to feed on toxic plants containing hundreds of chemical compounds that would kill a human. Other insects are remarkably tolerant of exposure to heavy metals, and even to poisonous chemicals specifically engineered to try to kill them. Hundreds of insect species have been reported to have evolved resistance to insecticides, and despite our best attempts to eradicate certain pest species over the past century, we have not exterminated a single one to extinction. Ironically, we can't seem to eliminate any of the ones we would really like to be rid of, like the malaria mosquito, the human body louse, the rat flea, or the house fly, while at the same time probably millions of nontarget tropical insect species may be immediately threatened with extinction by our unfortunate habit of sheer habitat destruction.

Perhaps it is easy to sound impressive by saying that there are more than one million insects, or anything else for that matter. Most of us don't own a million of anything, so in practice we don't count that high very often. But what really makes insect species diversity remarkable is not just the astronomically large number but the fact that we are talking about unique and different entities. To really grasp how extraordinary that is, one needs to begin with a clear understanding of what it means to be a species.

"And Whatever the Man called Every Living Creature— That Was Its Name"

In biology, the species is the most fundamental category for defining the kinds of living things. Since there are millions of different kinds of living organisms, you might not be surprised to learn that even biologists have a hard time coming up with a single definition for species. What works well for defining species of butterflies and beetles might not work as well for defining species of flowers, fungi, protozoa, and bacteria. Among the more popular ideas for defining species are the

biological species concept, the evolutionary species concept, the ecological species concept, and the morphological species concept.

The biological species concept defines species as populations of individuals that are able to interbreed and produce viable offspring and are
reproductively isolated from other such populations. In other words,
biological species consist of groups of individuals that will mate with
one another but will not normally interbreed with other species. This
concept works very nicely for most sexually reproducing insect populations, such as butterflies and bees. To cite a familiar example, the monarch butterfly (*Danaus plexippus*) is a very well-known and widely recognized insect species. The viceroy butterfly (*Limenitis archippus*), the
well-known mimic of the monarch, looks superficially similar in color
patterns but is a distinct and separate species. If you are patient and an
observant naturalist, you will see male monarch butterflies courting
and mating with female monarch butterflies, and you can observe male
viceroys courting and mating with female viceroys. However, you won't
find monarchs and viceroys interbreeding with each other or with any
other species, for that matter. The biological species concept attempts
to recognize and name the fundamental groups into which organisms
naturally segregate themselves. In that regard, the species category is
interesting, because it attempts to recognize groups that are not arbi-
trarily defined but have an underlying reality in nature.

The main problem with the biological species concept is that it
does not apply well to species that reproduce asexually, such as many
plants, fungi, bacteria, protozoa, and even some kinds of insects.
Many aphid species, for example, reproduce rapidly by having several generations of females that asexually produce more females without mating. Among the parasitic wasps there are many known species
where females simply produce more females by asexual reproduction
and males are totally unknown. The evolutionary species concept attempts to solve this issue by defining species as separate biological
lineages that share a unique evolutionary history and are genetically
distinct. As a theoretical concept this definition is more broadly ap-
plicable to all groups of organisms, but in practice it can be difficult
to apply. If we see male and female monarch butterflies mating, that
provides compelling evidence that we are observing two individuals
of the same biological species. Getting DNA samples from those same
two butterflies and assessing that they belong to the same evolutionary

species is still an expensive and challenging technological task. While our technology may be moving in this direction, the fact is that we only have assessed DNA "fingerprints" for a small fraction of insect species.

The ecological species concept defines species based on their ecological niches, that is, the unique combination of their habitat, feeding, environmental quality, and behavioral requirements. While the monarch and viceroy butterflies might at times occupy the same habitats in Canada, monarch caterpillars will feed only on milkweeds, while viceroy caterpillars will feed on willows, something a monarch would never do. The two species differ in their degree of cold tolerance and solve the problem in different ways, monarchs by migrating southward to Mexico, and viceroys overwintering as cold-tolerant, partly grown caterpillars. So the two species occupy different habitats at different times, and they utilize different resources for their development. A key part of the ecological species concept is the idea that no two species can occupy exactly the same ecological niche. Because they compete for living space and resources, species tend to diverge so that they adapt to use the world in slightly different ways. While this seems to provide a satisfying definition of how monarchs differ from viceroys, even the ecological species concept has a fundamental practical flaw: we don't know the ecological niches of many of the species that have been discovered.

Most named insect species were proposed based on the morphological appearance of collected specimens, size, color patterns, body form, and other distinctive anatomical characteristics. This brings us to the oldest and perhaps most fundamental definition: the morphological species concept, which characterizes morphospecies based on their anatomical appearance. This may seem old-fashioned or somehow less satisfying than the other species concepts, but in most cases it is extremely practical. I don't need to observe mating behavior, gather DNA evidence, or observe the larval food plants to tell the difference between a monarch and a viceroy butterfly. Just put a specimen in front of me, or even a photograph, and I'll tell you correctly which species it is, based only on its morphological appearance. Those two species each have unique and distinctive wing patterns, and people have been successfully recognizing monarchs and viceroys for more than two hundred years. Admittedly, there are some issues with the morphological species concept. Ranges of variation need to be assessed and understood, such as differences between sexes and variations between immature and adult stages. Also, we understand that in some cases there are such things as cryptic species that appear morphologically identical but can be differentiated by behavioral or genetic evidence. But the vast majority of living species can be defined based on their morphological appearance, and, as a practical matter, the species of most fossilized organisms can be defined based only on their morphology. This operational definition once prompted the paleontologist David Raup to remark, a bit cynically, that "a species is a species if a competent taxonomist says it is."[3]

While it is important to conceptualize what a biological species is in theory, it is also valuable for you to realize what a species is, in practice. For the past 250 years or so, biologists have been naming new species, and since 1961 this has been done according to various rules set forth in the International Code of Zoological Nomenclature. To describe and name a new insect species, the rules do not require you to have DNA samples, know the ecological niche or the evolutionary history, or even to observe mating biology. The code does require that you have a specimen, or part of a specimen, that can be observed and described and archived for reference in a museum collection. The actual process of naming a new insect species involves describing the morphological characteristics of the proposed new species, giving it a name, and publishing this information in a scientific journal; the date of publication is what makes the name official. Our system of naming species always uses binomial nomenclature requiring two words to state the full scientific name of a species: the first word is the genus name and second is the species name or epithet. Those two words form a unique combination, so that the species name for every living species is unique and distinctive. The species name is always Latinized but need not be complicated or difficult to learn (you probably already know your own species name, *Homo sapiens*). The specimen is kept in a museum collection for future reference, but for the most part, species become known by what is published about them in the scientific literature. So under the taxonomic species concept that is universally used for naming and discussing insects, a species is first defined as a set of organisms with a certain stated series of shared characteristics.

I prefer to think of naming new species as making a species hypothesis. When we define a species based on morphology, we are essentially hypothesizing that the same biological, evolutionary, and ecological species exists with that form. The species hypothesis is tested with each addition of new information. We hope and expect that in the future we will learn the biology, evolutionary history, and ecological niche of every named species, thereby corroborating the morphological species that have been proposed. With the discovery of new specimens, we gain new information about variations, and the taxonomic concept of an organism may be modified and expanded to include this new information. If the discovery of new information suggests that a named animal or plant is merely a population of some other species, then the older name is preserved and the more recent name is "sunk" and becomes a junior synonym of the older valid name. Or, if a population is discovered to consist of multiple cryptic species, then new names can be added to recognize the newly discovered biological species. Naming a new creature is just the first step in a long scientific process of developing a fuller understanding of the organism, but that process necessarily starts with simply providing a morphological description of the appearance of the beast. Having mentioned that scientists have been attempting to do this work for 250 years maybe makes it sound as if we might be nearly done with discovering and describing species. But keep in mind that given the current rates of new discoveries, description, and publication, it could take an estimated 500 more years just to provide names and morphological descriptions for the remaining insect species.

However you define them, there are at least several million insect species on this planet, and they all have one fundamental common feature: they are unique populations. It's not like having millions of the same thing. In some cases, sheer abundance of one insect species is amazing by itself. It's impressive to know that one ant supercolony might have millions of nearly identical workers. But it's mind-boggling to understand that this planet has millions of distinct insect species populations, each with its own unique reproductive biology, ecological niche requirements, genetic and evolutionary history, biochemistry, anatomy, and behavior. It's hard to conceptualize even the one million or so named insects. To visualize how many that really is we might look at it this way: since each species name consists of two words, it would require printing two million words just to state the names of the known insect species; I'd need twenty books about this size just to list all the insect names in print, in sequence. That's just the living modern insect species, not the extinct ones known from fossils or the millions of other unnamed ones living high up in tropical forest canopies. I can't hope to tell you something interesting and unusual about every single insect species, although every one certainly has a fascinating tale to tell. So, we need a system of grouping species together into meaningful categories for discussion.

Birds of a Feather, Butterflies of a Scale: Classifying Insects

Classification is the tool we have invented for assembling living diversity into hierarchical groups. Species are grouped together with other related species into genera. For example, if you live in eastern North America you are probably familiar with *Bombus pennsylvanicus*, the Pennsylvania bumblebee, while in my mountainous area of Wyoming the common big bee is Hunt's red-tailed bumblebee, *Bombus huntii*. These are distinct bee species, with different distributions, feeding habits, and nesting habits. Most people would generalize and simply recognize either one as a bumblebee, that is, a member of the genus *Bombus*. But even with this common example, the situation is far more complicated. We have about ten species of bumblebees in Wyoming, more than thirty in North America, and many more worldwide.

Groups of similar species and genera are classified into family groups. Bumblebees, honey bees, carpenter bees, and others are assembled into the bee family Apidae, consisting of hundreds of species worldwide. Among insects, family groups can vary quite a bit in size and diversity. One of the largest, the parasitic wasp family Ichneumonidae, is estimated to contain fifty thousand to sixty thousand species, which is more than all the kinds of vertebrate species combined. Several other insect families have comparable hyperdiversity, such as the Curculionidae (weevils), the Carabidae (ground beetles), the Noctuidae (noctuid moths), and the Tipulidae (crane flies). My research specialty is the parasitic wasp family Braconidae, which does not have a generally accepted common name, but consists of more than fifteen thousand named species and is estimated to contain well over fifty thousand total species. On the other hand, a family group can be very small: the wasp family Pelecinidae consists of just one wide-ranging species from Canada to Argentina.

Regardless of species diversity, groups of related families are assembled into larger groups called orders. The largest and commonest of the insect orders are very familiar, even if you don't know their scientific names. Four of the insect orders are hyperdiverse, each with thousands of described species and probably millions of existing unnamed species. These four biggest orders are the Coleoptera (beetles), the Hymenoptera (wasps), the Diptera (flies), and the Lepidoptera (butterflies and moths). There are many other insect orders that are comparatively smaller but still very large, consisting of hundreds or thousands of species, and you are probably familiar with many of them, such as dragonflies, grasshoppers and crickets, termites, roaches, lice, true bugs, and, especially if you like to fish, the mayflies and caddisflies. Lots of other interesting but rare insects are classified into small and obscure orders that most ordinary people never encounter, such as the webspinners, icebugs, zorapterans, and twisted-wing parasites. The smallest and most recently discovered order of insects is the Mantophasmatodea, also known as African rock crawlers and gladiator insects, with two living species named in 2002. There are also several orders of insects that are now extinct, but were formerly quite diverse in the Paleozoic era, hundreds of millions of years ago.

All these buggy orders are assembled into a much larger group called the class Insecta (the insects). For the most part, these are all the small creatures that people commonly call "bugs," creatures with six legs and three body parts: the head, thorax, and abdomen. The insects in turn are classified into a larger group called the phylum Arthropoda (arthropods). This includes not only the insects but also their near relatives with hard external skeletons and jointed legs, familiar creatures such as spiders, ticks, mites, millipedes, centipedes, and scorpions, as well as less familiar beasts like the extinct Paleozoic trilobites. Lastly, the arthropods are classified in the much larger kingdom Animalia (animals).

Six-Legged Menagerie: Understanding Insect Diversity

Why should one group of animals have become so astronomically diverse in comparison with all other kinds of animals? How are we to understand the ecological success of insects in particular? First, insects have evolved small body sizes, which are limited by aspects of

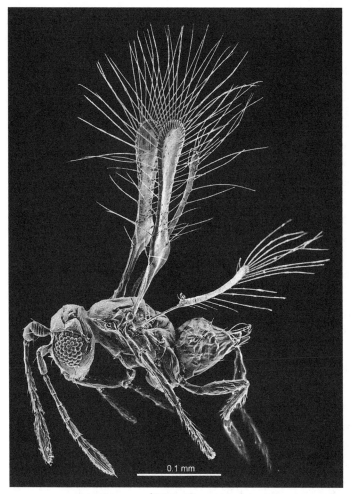

0.1 mm

FIGURE 1.2. Fairyfly wasps (family Mymaridae) are insect-egg parasitoids that occupy minute ecological niches. *Tinkerbella nana* is one of the smallest known flying insects. (Photo by Jennifer Read.)

their anatomy, physiology, and ecological interactions with other creatures. Small size has promoted insect species diversity by allowing bugs to divide the world into exceedingly small niches. Plant-feeding insects, for instance, may partition one plant into many different niches. You will find different kinds of insects feeding on leaves, boring in stems, mining under leaf surfaces, chewing in flower buds or seed pods, living under bark, or boring deep into the heartwood or below ground into roots. A young beetle may spend its entire imma-

FIGURE 1.3. Bat flies (family Streblidae) are highly specialized, blood-feeding bat parasites. This one lives on vampire bats at Palo Verde National Park in Costa Rica.

ture life living and feeding inside a single tiny plant seed, while the adult can live in a totally different niche.

While many common insects, such as the monarch butterfly, have a wide distribution and a fairly broad niche, the vast majority of insect species are extremely small and have very tiny, often localized niches. The tiniest insects, microscopic "fairyfly" wasps too small to see without magnification, develop inside the single egg of another insect. The specialized narrow niches of parasitic insects are also particularly impressive. There are bird lice that live only in the pouches of pelicans and mammal lice that live only in the fur and nostrils of sea lions. In fact, most different kinds of birds and mammals have specialized lice species living on their bodies, except for bats. But another group of insects has evolved to fill this ecological niche. Remarkably, there are bloodsucking parasitic flies that live in bat fur, so while vampire bats might drink your blood, their own blood is being fed on in turn. In Wyoming and other boreal parts of North America, there is a flightless parasitic beetle species, *Platypsyllus castoris*, the beaver parasite beetle, which lives only in the fur of living beavers. *Neoneurus mantis*, an insect I discovered in Wyoming, is a tiny wasp that develops parasitically inside the abdomens of one species of mound-building *Formica* ant. When I first found them in 1990, they were flying in association with only three particular ant colonies. Over the past two decades, two of those colonies have declined in numbers, and the parasites have disappeared. So over the past several years, I've only been able to find this tiny wasp in the nearby mountains in June, at only one particular ant nest.

A second reason for their vast diversity is flight, which has allowed winged insects to expand their niches into the air. For as long as 150 million years, insects were the only animals that could fly, and that gave them great advantages in terms of their ability to escape predators and to disperse and colonize new areas. The eventual evolution of winged pterosaurs, birds, and bats didn't drive insects out of the air. It simply provided the selective forces to drive additional diversification and specialization. But wings are also very useful for things other than flight. Insects are cold-blooded animals, and since blood flows through the veins of their wings, insect wings work very well as little solar panels for warming up on cold mornings. Insect wings are also canvasses for multifarious colors and patterns that play important roles in insect behavior, especially in courtship and mating, and facilitating mate recognition in diverse and complex environments. Wing colors can also enhance survival by either crypsis (camouflage) or aposematism (bright warning coloration).

Third, but certainly not least, insects have evolved elaborate methods of development, including complex metamorphosis with young developmental forms (larvae) that are stunningly different from adults. That was a remarkable innovation, which allowed adult insects to avoid competing with their own offspring for food. A young insect larva, such as a caterpillar of a butterfly, becomes a feeding machine that can concentrate its activities on eating and growing, often in a concealed or protected environment. In contrast, adult insects with complete metamorphosis can concentrate their activities on courtship, mating, and egg production. Many feed on completely different resources from their young, or do not feed at all. The fact that more than 75 percent of all modern insect species have complete metamorphosis attests to the usefulness and success of this behavior.

An inquiring student of insects might seek more sophisticated answers in the study of insect evolution over the past four hundred million years. An advanced course on this topic might cover the history of insects based on fossil remains, and relationships inferred by studies of anatomy, behavior, genetics, and molecular sequences. One would probably learn that the most ancient insects evolved on land in association with the most ancient land plants, in the Late Silurian or Early Devonian periods. By the Carboniferous period, the time of the great coal-forming swamps, insects had evolved wings. The innovation of flight allowed the first great radiation of insect species, including giant dragonfly-like insects and many other ancient species that no longer exist. By the Permian, insects were diversifying into increasingly modern forms, including species we would recognize as bugs, lacewings, and beetles. The development of complete metamorphosis in the Permian was a key innovation, perhaps, that allowed many insects to thrive and diversify in an unstable and changing world. Other insects, and many other creatures, were not so lucky.

At the end of the Permian, our world experienced a massive extinction episode. In the oceans many ancient life forms, including the trilobites, disappeared for all time. On land several kinds of life went extinct as well, including some of the ancient insects that had flourished for millions of years. But others survived and diversified, especially those with complex metamorphosis. And so it was through the Mesozoic times, when the insects coexisted with the large dinosaurs. Sometime in the Early Cretaceous period flowering plants evolved, and insects adapted by rapidly evolving diverse new species. Along with the flowers came the insect associates: butterflies, bees, ants, and social wasps. Another major extinction event, perhaps triggered by a massive asteroid impact sixty-five million years ago, brought to a close the time of the big dinosaurs. The little feathered dinosaurs, which we now call birds, survived and flourished, as did many flowering plants and insect groups. And so it has gone through the Cenozoic: the ongoing coevolution of flowering plants and insects has generated awesome biological diversity and complexity in tropical forest ecosystems.

The story of insect evolution, as revealed by their rich fossil history, though illuminating, is not completely satisfying. We might still wonder about the origins of the first insect. I began this chapter with a quote from Barry Hughart, who pointed out that "all events have an end and a beginning." To gain a full understanding of something as complex as insect diversity on earth, we need to "understand correctly what comes first and what follows." We need to delve even deeper into the insects' prehistory, into the time before the Silurian myriapods heroically occupied the land, with their external skeletons and legs. The myriapods inherited these features from sea-dwelling Cambrian arthropods, which first developed body armor and mobility in response to increasing oxygen levels and predator activity in the oceans. So our story of insects must begin there in the earth's ancient oceans, more than a half-billion years ago.

2 Rise of the Arthropods

About 600 million years ago . . . all hell broke loose in organic evolution.
DAVID RAUP, *Extinction: Bad Genes or Bad Luck?*

Sometimes the destination isn't as important as the drive.
CHRISTINA APPLEGATE, *Up All Night*

If you want to experience the geological ages of animal life on earth, then I highly recommend the scenic drive from Shoshoni to Thermopolis, through the awesome Wind River Canyon in Wyoming. The exposed rocks are a window on the past opened by the formation of the Rocky Mountains. As crust was thrust upward it tilted on its side, depositing the oldest rocks at the top of the canyon. So the drive from Shoshoni initially follows a path of Precambrian rocks devoid of animal fossils. A road sign near the canyon's entrance marks the start of rocks from the Cambrian period, when life finally showed the inclination to crawl out of the bacterial slime, and the first animals with skeletons evolved. It was the time of the first arthropods, and is especially well known for the rapid diversification of the ones commonly known as trilobites. The scenery is stunning. This is the only highway I've ever seen where the ages of animal life are marked by signs.

In a half-hour or so, you drive through all the ages. After passing through the Cambrian, the "age of invertebrates," you enter Ordovician time, the "age of fishes." Soon you are at the Silurian age, the time of the first land plants. Then on to the Devonian time, the so-called age of amphibians. The limestone layers of Silurian and Devonian time suffered extensive erosion in this area of Wyoming, so these layers pass in the blink of an eye. They are followed by the Carboniferous, the time of coal-forming swamps. Next you are on to the Permian, and as you approach the foothills and exit the canyon, you leave the Paleozoic era and enter the Mesozoic, the "age of reptiles," or as we now say, the "age of the dinosaurs." Take a brief moment to mourn the passing of the trilobites. But at the same time, you can rejoice in the appearance of the dinosaurs. They are celebrated at the end of the drive at the excellent Dinosaur Center in Thermopolis. Be sure to budget enough time to soak in the sulfurous hot springs teeming with bacteria and to also visit that impressive museum. Finally, as the road levels and the sagebrush prairies open to the valley, you enter Tertiary times, the first years of the Cenozoic era and the onset of the "age of mammals." Lament, if you must, the passing of the big dinosaurs. But rejoice in their passing as well, because it finally, finally allowed us mammals to make our move on this planet.

By now, the implicit human-centrist bias in some of that history I just recounted should be obvious. By noting the times of the proliferation of mammals, reptiles, amphibians, and fish, we are merely observing some tenuous history of events leading to the origin of the human species. How ridiculously unlikely it seems that we should be here at all, and how those labels for the geological periods distract from the real pattern of life's diversity. I'll be picking apart this human-centrist mythology bit by bit as we proceed. For now, it will suffice to examine the Cambrian.

The Cambrian period has generally been called the "age of invertebrates." That's certainly not because anyone sought to glorify our invertebrate ancestry. It's simply an observation that we didn't initially see any of our vertebrate ancestors in fossils from Cambrian layers. Notice that we didn't call it the "age of arthropods" or the "age of trilobites," either of which would be apt. Calling it the "age of invertebrates" is a bit like calling it the "age of no humans." The name subtly derides the success of arthropods by noting the absence of vertebrae rather than touting the evolution of exoskeletons. But subsequently, we did discover our likely vertebrate ancestor in Cambrian times, and what a humbling event that was. A small creature called *Pikaia* was discovered in the 515-million-year-old Burgess Shale fossils of Canada. *Pikaia* was a mere one-and-a-half-inch-long, wormlike creature that burrowed in bottom sediments. She was soft-bodied but did have an internal supporting structure: a primitive notochord, the ancestral structure of a vertebral column. *Pikaia* is now regarded as the most likely common ancestor of fish, amphibians, reptiles, dinosaurs, birds, and mammals. But she was such a modest ancestor that no one lobbied for a renaming of the Cambrian as the "age of *Pikaia*."

When viewed from a great temporal distance and with a nonhuman eye, the history of life appears differently. It's certainly clear that there was no inevitable or rapid progression toward humans. A nonhuman space traveler might rewrite our biological history far more succinctly. The first three billion years or so might simply be called the "age of bacteria." The time from the Cambrian to the present (the last half-billion years or so), the time of multicellular animal life, could simply be called the "age of arthropods." Since the onset of animal complexity, the arthropods have been the singular successful group, both in diversity and abundance. The rise of insect diversity was ancient enough that the last three hundred million years or so could be dubbed the "age of insects." The last ten thousand years, during which human civilizations have arisen, is but a miniscule and insignificant blip compared to the vast spans during which first bacteria, then arthropods, then insects in particular, have dominated this planet's landscapes.

The other remarkable thing about the Cambrian is simply that *we* survived it at all, and by "we" I mean not just humans but the entire lineage of vertebrate animals. There it was, *Pikaia*, our humble worm-like ancestor, tunneling along in bottom sediments. It is quite rare in Cambrian fossils and perhaps was never a very abundant creature, even in those days. In the waters above, along cruised animals like *Anomalocaris*, a three-foot-long, nightmarish predatory arthropod with long, spiny feeding appendages. *Anomalocaris* paddled along in the Cambrian seas, picking off whatever small animals it could catch— no doubt feasting on lots of trilobites. From time to time, *Anomalocaris* doubtless swooped down to pick off a tender *Pikaia* for dinner. Luckily for us humans, most of the abundant Cambrian arthropods, trilobites, also fed in the bottom sediments, peacefully, along with *Pikaia*; however, there is evidence that some of the Cambrian trilobites may have been predatory. There are fossils of trilobite tracks intersecting with worm burrows and resting, which suggests that some trilobites may have preyed on soft worms in the sediments. Trilobite body forms and mouthpart styles are certainly diverse enough to suggest that they had evolved a comparable variety of feeding habits. But still, if Cambrian trilobites had become extensively predatory, then it's exceedingly unlikely that *we* would be here to piece together this story.

Life's a Blast: The Cambrian Explosion

The Cambrian period, roughly 541 to 485 million years ago, was a definitive time in the development of life on earth. Finally, after 3 billion of years of microbial history, single cells assembled into functional groups, and multicellular animals appeared. Animals wasted little time, geologically speaking, in evolving structural support and protective gear: cuticles, skeletons, and shells appeared over a period of only 5 million years. The seemingly rapid evolution of early animals has prompted paleontologists to dub this event "the Cambrian explosion" of life, and this time is notable for the rapid evolution of diverse phyla, including the major lineages of organisms that still dominate the planet. Most relevant to our story is the first appearance of the phylum Arthropoda, the armor-plated lineage from which the insects would eventually emerge.

With the evolution of hard parts in animals, the earth's geology was forever changed. Those hard parts fossilize well, and while fossils of Precambrian soft-bodied organisms are rare, the remains of hard-bodied Cambrian critters left comparatively abundant fossils. So from the Cambrian onward, the earth's rock layers literally preserve impressions, pressed snapshots of past life.

These rock layers have been accurately dated using radioactive isotope distributions, but those from the Cambrian onward can be easily identified by the kinds of fossils found in them. The signature fossil group of that period, no doubt, is the trilobite. These small creatures had a segmented, three-lobed hard skeleton, hence the name: they are trilobed. Trilobites are some of the first common examples of the larger animal group into which insects are also assigned: the arthropods. Insects, spiders, lobsters, shrimp, millipedes, centipedes, scorpions, and trilobites all share the basic arthropod anatomical innovations: a hard, segmented external skeleton and several jointed legs. We will learn more about that later. For the moment, I just want you to recognize that those skeletons, being hard structural parts, fossilized very well. Trilobites were abundant enough in ancient shallow seas that their skeletons were frequently covered by sediments. Nowadays, you can take a walk in the prairies of Wyoming's Bighorn Basin and see that the rocky landscape is studded with trilobite fossils. It's a sure sign that the area was once under ocean water, and the rocks are of Paleo-

FIGURE 2.1. The fossilized molted exoskeleton of a Cambrian trilobite, *Elrathia kingii*, a common species in the Wheeler Formation of Millard County, Utah.

zoic age. However, you won't find trilobites or any other animal fossils in rocks that are 3 billion years old. They went extinct by around 252 million years ago. So you won't find them in rock layers from the middle Mesozoic, mixed with dinosaur bones. And you certainly won't find trilobites in places like Hawaii and San Ramon, Costa Rica, because those landscapes were formed by volcanic activity over the past few million years. They have no rocks dating back to the Paleozoic.

But in the quarries of the continental United States there are lots of Cambrian-aged rocks, and trilobite fossils are fairly common. You don't need to be a paleontologist to see one. Walk into any rock shop in North America and you will probably find a bin full of trilobite fossils, most likely with a sign declaring them to be the oldest of animals. So common are the trilobites that you can buy a trilobite fossil for a couple of bucks and carry it around in your pocket, if you wish. In Ohio, Pennsylvania, and Wisconsin, they are official state fossils.

Skips in the Fossil Record

As much as we can learn by examining fossils, it is important to remember that they seldom tell the entire story: the fossil record is never as complete as we would wish. Things only fossilize under certain sets of conditions. Shallow marine communities with frequent sedimentation produce fossils comparatively well, so the record of Cambrian animals is not so bad. Modern insect communities are highly diverse in tropical forests, but the recent fossil record captures little of that diversity. Many creatures are consumed entirely or decompose rapidly when they die, so there may be no fossil record at all for important groups. It's a bit similar to a family photo album. Maybe when you were born your parents bought a camera and took lots of pictures, but over the years they took photographs sporadically, and sometimes they got busy and forgot to take pictures at all. Very few of us have a complete photo record of our entire life. Fossils are just like that. Sometimes you get very clear pictures of the past, while at other times there are big gaps, and you need to notice what they are. One example from the Cambrian should suffice to make the point. A microscopic fossil of a tardigrade, a cute little animal that looks rather like a miniature teddy bear has been found in Sweden's Cambrian sediments. Tardigrades, also known as water bears, still exist today—we have discovered them in water samples from bromeliad tanks in Ecuadorian forests—but no intervening fossils have been found. The fact that they were found in two places, the Cambrian shallow marine communities and the wet forests of the modern world, doesn't mean that water bears evolved twice. It does show us that these animals evolved as early as the Cambrian and have persisted since, although there is no fossil record of it.

Rocks over time do not show a record of gradual steady change, and they certainly do not show any rapid progression toward humans. What the layers do show is a record of distinct times when communities of life emerged, and remained stable for millions or tens of millions of years. The reason we've divided the ages of life into these various times is not because we just wanted to divide it up. It's because the layers themselves present distinct communities of life, and the interfaces of layers show sudden rapid change, often after tens of millions of years of stasis. When times are good, life doesn't evolve simply because it has the capacity to change. Instead, when times are pleasant, species and communities of life tend to get into an equilibrium mode, where they're well adapted to existing environmental conditions and move along comfortably for long periods of time. But the record of life is punctuated by occasional dramatic events that really give life a jolt: glaciers, continental drift, comets, asteroids, and the like. Occasionally, communities of life have suffered catastrophic mass extinction. But each time change occurs new species quickly evolve to fill the empty niches until a new equilibrium of life is established. This is what biologists call "punctuated equilibrium," a term that was coined by the paleontologists Niles Eldredge and Stephen Jay Gould.

Setting the Stage for Arthropods: What Ignited the Cambrian Explosion?

One of the most striking observations about the history of life is the significant fact that life on this planet remained single-celled for roughly three billion years. Why did it take so agonizingly long for multicellular animals to evolve? A simple answer has been suggested: oxygen.

For 3 billion years, ancient bacteria bubbled away, making oxygen and binding carbon dioxide into calcium carbonate and carbon-based sediments. They may have been releasing oxygen, but for a long time the oxygen content of the atmosphere did not change much. Before oxygen could accumulate in the air, the free oxygen reacted with iron and other substances in the earth's crust and oceans. It was locked up into sedimentary rocks, banded iron formations, and minerals for millions and billions of years. On the other hand, massive amounts of available carbon from excess carbon dioxide in the atmosphere were being drawn down and locked up into limestone deposits. During that period of mineral formation, a fortuitous balance was set between the sun and the earth. The ancient sun was cooler, but the thick carbon dioxide-rich atmosphere of old earth provided a warming blanket in the form of a greenhouse effect. Over time, as excess carbon dioxide was drawn down out of the earth's atmosphere by living processes, the sun was growing warmer, so the earth remained comfortable for life. But eventually, around 2.3 billion years ago, the carbon dioxide level dropped too low, the earth entered a catastrophic ice age, and life experienced the first of what would be many punctuating events.

After billions of years of floating in balmy seas full of tasty amino acids, global glaciers overwhelmed the oceans, and most of earth's ancient microbial life was presumably exterminated. Survivors probably included any bacteria lucky enough to be at comfortable interfaces where volcanic vents met and intermingled with frigid waters. Today, these kinds of bacteria thrive in Antarctica and in Yellowstone's steaming volcanic vents and sulfurous hot springs. It was a tough situation for life, to be sure, but one that isolated cells into unique microenvironments and allowed genetic lines time to drift apart; natural selection insured that the remaining cells were indeed real survivors.

The global ice age, which may have lasted for millions of years, is indicated in the earth's rocks by banded iron layers; these layers formed when iron accumulated in the oceans then precipitated into sediments. The rocks are capped with a calcium carbonate layer, indicating that the ice age ended abruptly with a period of global warming when continental minerals were washed into the seas, stimulating a worldwide flush of bacterial growth. Oxygen was ejected back into the atmosphere, and the world teemed once again with eukaryotic cells.

We might suppose that such a near-death experience might have been the necessary jolt to set life along a more complex pathway. But that does not seem to be the case. Bacterial cells resumed their old pattern of floating around for tens of millions of years. Then, about 850 million years ago, continental drift brought the land masses into an unfavorable configuration near the equator, and again the earth was cast into a planetary deep freeze. Glacial ice approached the equator, and the chill lasted for millions of years. Finally, the ice was broken, and life enjoyed a brief reprieve. But this time a cycle was established, and between 850 and 590 million years ago the earth experienced not just one but at least four global ice ages. The most recent of these, the great Varanger ice age, lasted 20 million years, from 610 to 590 million years ago; scientists have dubbed it the time of the "snowball earth."

As the last snowball earth came to an end and the last global glaciers retreated toward the poles, life reassembled into multicellular clusters. Soon after that, abundant early animals appeared in the "Cambrian explosion." What finally stimulated such dramatic changes in life forms, after billions of years of single-cellular domination? What happened most notably is that atmospheric oxygen levels finally rose to levels approximating our modern atmosphere. Potential oxygen toxicity drove cells into clusters for safety, but at the same time an energetic system existed to motivate animal life: aerobic respiration. Animals live more complex and energetic lives than bacteria because oxygen forced them to do it, and oxygen enabled them to do it. This process of oxygen levels affecting the evolution of life, the history of changes in oxygen levels, and the geological evidence for all this is thoroughly covered in Nick Lane's popular and entertaining book, *Oxygen, The Molecule That Made the World*. So, there's no need for me to repeat it all here.

After the last snowball earth, another very important event happened: continental drift accelerated and the Precambrian continents were dramatically reconfigured. The continental drift rate during the Cambrian has been estimated to be about ten times faster than the average rate since then. This was important for two reasons. First, it brought continental land masses back near the poles in a surprisingly rapid fashion. This stabilized the planet by reestablishing the cycle of weather that keeps ice ages more moderate. Some land masses have been near a pole, one way or the other, since that time, keeping modern earth out of the severe "snowball" phase. Second, for the Cambrian animals it was a bonanza, because rising sea levels and rapidly drifting continents meant lots of shorelines and more shallow marine communities with abundant mineral sediments. The earth was spinning faster during the Cambrian and the moon was closer, so tidal forces on shallow marine communities caused rapid pulses of nutrient flow. The time was ripe for rapid evolution of animal life.

Skeletons in the Cambrian Closet

Some of the earliest Cambrian animal action took place in shallow marine sediments. Among the oldest Cambrian fossils are "trace fossils" that do not show the actual animal, but animal tracks. These are abundant fossilized burrows, presumed to be caused by ancient marine worms. The oldest Cambrian animal to be given a name is *Tri-*

chophycus pedum, based entirely on fossil burrows. It precedes the appearance of any hard-shell fossils. Having only the tunnels, we don't know for sure if *Trichophycus* was one species of animal, or several with similar habits. Nevertheless, we can deduce a surprising number of things about them. Since they lived in bottom sediments, we assume they fed on accumulating organic sediments from bacterial life. The tunnels are long, narrow, and directional, so we know they could burrow through the sediment. This implies that they had front and back ends, a mouth and anus, and a digestive system. Since they moved and made tunnels, they must have had muscles and therefore an opposable cuticle system, some method of circulation, and of course, respiration. Most likely *Trichophycus* was a segmented worm-like creature, similar to the annelid worms, and insects presumably have inherited their segmentation from such ancestors.

The next layer of the Cambrian rocks is the first to contain hard part remnants of animals: tiny shells, spines, and small hard pieces that could be traces of the earliest external skeletons and are difficult to assign to more modern groups with any certainty. This layer is called simply the "small shelly fossils" or the "early shelly fossils." While not much is known about these ancient animals, they do teach us something important. You may recall that the ancient cyanobacteria secreted calcium carbonate to form stromatolites. The evolution of external shells was a similar process. The early shelly animals built portable hard parts simply by secreting waste products that solidified. As animals evolved predatory habits, the aspect of shelliness would have immediate benefits. Aside from a protective covering, hard parts form the basis for skeletal systems, providing the opposable parts for musculature. So the evolution of shells was a step toward the evolution of more complex skeletons, musculature, and ultimately faster locomotion.

In the next layers of Cambrian rocks, layers 529 million years old and more recent, we begin to find the fossils of the so-called Cambrian macrofauna. These are the first fossils of animals with full skeletons and distinct limbs, which became more abundant as the Cambrian elapsed. For the most part, they are fossils of trilobites, and other arthropods. Examples of other recognizable groups (other phyla), such as sponges, corals, mollusks, and annelid worms also are present. There were also a bunch of weird and wonderful animals, unlike anything modern, that lived for a while then disappeared.

Often you will hear that most of the modern animal phyla appeared in the Cambrian period. This means only that we see the first examples of arthropods, annelids, mollusks, echinoderms, and chordates, the ancestors from which modern groups can trace their lineages. It certainly doesn't mean that animal groups burst on the scene with anything like the species diversity that exists in modern animal phyla. It simply means that during the Cambrian the first arthropod species appeared, the first mollusk, the first echinoderm, the first annelid, the first chordate, and that each of these groups developed the basic body plans which characterize the modern animal phyla we see today.

Rock Stars of the Cambrian Seas

The real success story of animal diversity is the arthropods, as exemplified by the Cambrian trilobites. Found in the oldest Cambrian layers with the first Cambrian macrofauna, trilobites lived in the oceans until the end of the Paleozoic era, diversifying over a span of nearly three hundred million years. We have discovered nearly twenty thousand species of trilobites, most of which lived in the Cambrian and Early Ordovician. By the Late Cambrian, trilobite diversity peaked with more than six thousand species classified into eight hundred genera and seventy different trilobite family groups. Most of the trilobites dwelled on or burrowed into the shallow bottom sediments. Some large bottom-dwelling species appear to have had immature forms that were planktonic. But some trilobites could swim, and other small species appear to have been planktonic as well, moving about with the currents and tides. The trilobites may be long gone, but all modern arthropods have inherited some similar aspects of their body form: namely, a hardened external skeleton and several multijointed legs.

Let's consider the evolution of skeletons, because if anything, the Cambrian explosion was a proliferation of hard parts, an explosion of skeletal diversity. Much of what's been said about Cambrian animals in the popular press has focused on the weirdness of these animals' skeletal forms. The Cambrian menagerie included strange creatures like *Hallucigenia*, which was so spiny and leggy that for years we didn't know which side was the bottom or the top, or which was the head or the tail. But let's not get distracted by the weirdness of subsequent modifications. When skeletons first evolved, there were only a couple possible approaches. You could build your skeleton on the outside, supporting and protecting your soft growing cells on the inside. Or, you could build a skeleton on the inside, purely for support. Basically, you could have either an external skeleton (trilobite style) or an internal skeleton (fish style). Some animals with internal skeletons might also mimic the arthropod anatomy by adding some exterior body armor: armored fishes, plated dinosaurs, modern-day armadillos, and King Arthur's knights. But fundamentally, skeletons come in two styles: outside and inside. Both skeletal styles provide the necessary structural support for muscle attachment and locomotion, a key aspect of what it means to be an animal.

The advantage to having an external skeleton should be immediately obvious: it provides protection as well as support. It's the same reason why we mostly wear shoes. The disadvantages are more subtle: an outer skeleton places some limits on sensory systems, as well as limiting growth. Arthropods may not have an outer skin to feel things as we do, but they compensated by covering the skeleton with sensory spines. Growth is more challenging for an arthropod; it's tough to keep growing when you live inside a suit of armor. This required arthropods to evolve metabolic pathways allowing them to periodically molt an old skeleton and regrow a new one. It's an adequate solution, but it does mean that they all have times when they are temporarily soft-bodied and vulnerable, like a soft-shelled crab. No doubt that tender stage is the most vulnerable to predators. We vertebrates, with our tedious internal skeletons, have only one real advantage over the arthropods. We do not need to molt or regrow new skeletons. Our growth period can be continuous, without such interruptions. Of course, we have the very serious disadvantage that our tender, tasty outside is constantly exposed to predators and the environment. So to protect their exposed bodies, vertebrates have compensated with scales, slime, armor plates, feathers, fur, and Levi's denim. We protect our tasty flesh by encasing ourselves in the hard metallic shells of massive fossil-burning automobiles.

I present to you a simple proposition: when it comes to animal form, an external skeleton is better. It's certainly more likely to develop in the first place, because it's better to shunt toxic excretory byproducts to the outside rather than storing them on the inside, and it's obviously far more likely to succeed because of immediate defensive advantages. Just look at the vast diversity of trilobite species that lived at the end of the Cambrian period versus our little *Pikaia* relative hiding in the sediments. Then also consider our modern world, where all vertebrate species combined present only a small fraction of seemingly astronomical arthropod diversity. Success might be measured in various ways. We are ever so proud of our oversized contemplative brains. But with those brains we must ponder the sheer improbability of our existence and our constant vulnerability through the ages. The humble insects cannot contemplate their own measure of success: the numerical dominance of arthropod species through all ages of animal life. If this were to all play out again on another planet, it seems to me highly improbable that soft-bodied creatures with internal skeletons would develop first or become successful over the long run. Hard-covered creatures with external skeletons would almost certainly hold the advantage over time, in most contests of soft versus hard-shelled players.

Looking back on the earliest Cambrian trace fossils, those worm burrows, we know that even those simple animals could tunnel in sediments. We don't suppose that they had any legs yet, because there are no fossil footprints. So how did they move? We must assume they had muscles arranged in body segments, allowing them to contract segments and wiggle their bodies, as with modern earthworms. Segmentation is a common body form, but it is an ancient one as well. All modern insects are segmented animals; hence the name "insect," which means "in sections." But that characteristic is not unique to insects. It is an inherited trait from earlier ancestors. All arthropods, including the trilobites, were segmented creatures, and so were simpler creatures, such as the annelid worms. That ancient burrowing creature *Trichophycus pedum* was probably also segmented, precisely because it could tunnel, but did not have apparent legs.

The origins of segmentation clearly reside in the earliest multicellular animals. Just as single cells became multicellular aggregations by building duplicates of themselves, early multicellular creatures became segmental by building duplicates of their cellular arrangements and linking them in a chain. Segmentation is an excellent trait for an animal, not just because the components are easiest to build and link together, but also because the shock is less when you lose a part to an accident. If you have seen the science-fiction movie *The Core*, you may recall the following example: explorers used a multisegmented craft to travel deep into the earth, losing parts along the way but surviving. Trilobites and insects can lose body parts and survive more easily than we can.

It's no mistake that the earliest skeletal parts were on the outside. As I already mentioned, it makes more sense to shunt waste products to the exterior than to pile them up on the inside. If an ancient worm-like creature were evolving a hard outer skeleton, it is only logical that the skeleton would form in segmental plates, just as in all arthropods. They already had segments with muscles. Body flexibility could only be maintained if the segmental parts remained flexible, with membranes at the edges. Any attempt at a fully hardened exterior would be useless and maladaptive, because a completely hardened creature could not move at all.

The other unique feature of arthropods, and another key to their success, is their multijointed legs. Take a moment to imagine a delicious plate of steaming Alaskan king crab legs, and you know what I'm talking about. Jointed legs in hard shells, that's the feature that defines the arthropods and the characteristic was passed along to the insects. The name "arthropod" translates to "jointed foot." Here's an easy way to remember that. You recall that when you have problems with your joints, we call it arthritis. For problems with your foot, you see a podiatrist. So, arthropods have a jointed foot.

Among the Cambrian fossils there are several indications of leg origins. Some Early Cambrian annelid worms like *Burgessochaeta* have short paired protuberances on each segment. These appendages were not jointed but had long bristles. Clearly they were not legs but could still have been useful for locomotion. Even in the simplest of worms, just the smallest pair of segmental protuberances would have improved friction with the substrate, and would have been useful for burrowing. Appendages appeared in pairs, two per segment, simply because most of the Cambrian animals possess bilateral symmetry. Slice one lengthwise and you get two similar halves, like mirror images.

Jointed legs appear not just in trilobites, but also in an assortment of other Cambrian arthropods, some of them still unnamed. There were multisegmented, multilegged creatures that resemble millipedes or centipedes but lived in the oceans. There were small-shelled arthro-

FIGURE 2.2. A white-legged millipede illustrates some character-istics of arthropods: an external skeleton, segmentation, and paired, jointed legs. (Photo by Kevin Murphy.)

pods that appear to be early crustaceans, the forerunners of lobster and shrimp. Also, there was the four-inch-long, armor-plated *Sanctacaris*, possibly the aquatic prototype that led to scorpions and spiders. But there were no insects, not yet. How did any of these things get jointed legs? Presumably in the same way the long skinny animals gained segmentation. A long, non-jointed leg is inflexible and limited in its usefulness. Any arrangement of leg joints, however, is very useful, allowing flexibility and the ability to manipulate potential food objects. As the history of the arthropods demonstrates, this simple leg form can be easily modified into an astonishing array of forms and functions. Insects use their legs for walking, running, hopping, fighting, grasping food, tasting food, grooming their body, swimming, digging, spinning silk, courtship, sound-production, and even hearing. Katydids have ears on their legs.

The Rise and Fall of the House of the Trilobites

If arthropods are so great, then what happened with the trilobites? Understanding the fate of the trilobites will explain not only why the realm of the trilobites rose and fell. It also reveals something fundamentally important about arthropod biology that helps to explain the later rise and success of the insects, which, after all, are the trilobites' distant relatives. They are not derived directly from trilobite-ancestors but are more like distant cousins. As we will see, insects succeeded partly by solving some problems that the trilobites were never able to master.

I mentioned earlier that trilobite species diversity increased steadily during the Cambrian, peaking at the end of that age. At the onset of the Ordovician period, trilobite diversity started to decline, and it continued to drop until the Silurian. By the end of the Permian, they were all extinct. What caused the trilobites' decline?

The start of the Ordovician was marked not by catastrophic environmental events but by significant changes in the communities of living organisms. New and different kinds of organisms appeared in the shallow oceans, and many of the more unusual Cambrian animals, like *Hallucigenia*, disappeared forever. From our perspective, the Ordovician most notably marks the time of the appearance and beginning of the diversification of fishes, the first obvious vertebrates with extensive skeletal features. For that reason, the Ordovician is usually highlighted as the time of the first fishes.

Once again we need to tear down some human-centrist mythology. We acknowledge the first fishes not because they dominated the animal communities; instead, we hail them as our most ancient vertebrate ancestors and assume that their appearance must have been a historical event from a human perspective. The truth of the matter is that fishes didn't change things too much and not very quickly. And fishes were not the most diverse or dominant animal group in the Ordovician. Trilobites started to decline in diversity over the Ordovician, but they still vastly outnumbered fishes for those sixty-two million years. By the middle Ordovician the trilobites declined to about thirty-five families, but there were still only five families of fishes. At the same time there were about fifty families of cephalopods, large predatory squids with coiled shells. We might well have combined the Ordovician time with the Cambrian and called it all the "age of trilobites." Or if you want to reflect the changes in the biological communities, then we should call it the "age of the cephalopods." But calling attention to the appearance of the first fishes is just a manifestation of our egos. If you looked at Early Ordovician communities, you would still see lots and lots of trilobites but not many fishes.

What really happened then is that some descendents of little *Pikaia* sprouted gills and fins and started swimming. Several million years spanned between the early *Pikaia* fossils and fishes' appearance, so we must suppose that however vulnerable that modest little creature may have been, she still had established a way of life that allowed her to survive for millions of years among the Late Cambrian arthropod communities. Then, at the start of the Ordovician, and with the appearance of fishes, there is no longer any trace of little *Pikaia*. Never again do we see soft little worms with notochords or vertebrae. Maybe the lineage of *Pikaia* was transformed entirely into fishes. Or, perhaps, those descendents that became fishes turned right around and gobbled up the last *Pikaia*.

Since the start of the decline of the trilobites coincides with the appearance of fishes, it's tempting to think that fishes played an important role in their disappearance. Probably they were a contributing factor, but it's obvious that nothing changed very quickly. Trilobite diversity started to decline after the Cambrian, but they didn't disappear entirely until the end of the Permian, about 250 million years later. Fish diversity increased only slowly and at first the fish were jawless, no doubt nibbling in the sediments. It took tens of millions of years before fish developed bone-crushing teeth to crack shells. The trilobites retained their advantage of hard external skeletons, and they were vulnerable to predators only during their small planktonic stages, and during their soft-shelled molting phase. Also, from the Ordovician to the Permian times, while trilobites were in gradual decline, the family diversity of cephalopod squids was always much greater than the diversity of fishes, and marine crustaceans were on the rise. Another predatory arthropod group made its first appearance in the Late Cambrian, the eurypterid sea scorpions. These nasty creatures had large armor-crushing claws, so they probably could consume large trilobites far more efficiently than either fish or cephalopods, which may have consumed more small planktonic forms of the trilobites. Finally, even the trilobites may have contributed to their own demise by evolving predatory species. During the Ordovician time there appeared the giant predatory trilobite *Isotelus*, which reached sizes of 16 to 28 inches. This ferocious monster has been dubbed the "*Tyrannosaurus of the Ordovician*" and has been declared the state fossil of Ohio. So although increasing predation pressure was a perhaps factor in the decline of the trilobites, it is clear that the fishes were not the main reason for their decline, and probably fishes were not the dominant trilobite predators.

To truly understand the disappearance of the trilobites, we need to look beyond the predators and examine some aspects of trilobite biology. Danita Brandt, a trilobite biologist at Michigan State University, has been studying the molting processes of trilobites, and she may have uncovered an important clue regarding their decline. Ironically, it seems that the key to trilobite's initial success, their external skeleton, may also hold the secret of their demise. It seems that although trilobites were among the first to successfully develop an arthropod skeleton, they never fully perfected the process of living inside it.

Danita Brandt has also been examining fossils of immature trilobites that died during the molting process. She found that the trilobites had a very irregular and inefficient method of molting their skeleton. Modern arthropods, like insects, have mastered the art of escaping from their skeletons by evolving an ecdysal suture: a line of weakness along the upper side that allows them to "unzip" the old skeleton. Insects do not molt until they have extensively recycled materials from the old skeleton and built a flexible new skeleton underneath. With the insects, the new skeleton hardens quickly, enabling them to regain normal functions within a matter of hours. Trilobites, by comparison, lacked both of these innovations. The process of breaking and escaping the old skeleton was irregular, even within a particular species it happened in various ways, and there were many deaths during this process. Even more troublesome for the trilobites was the extended vulnerable period between molts. It seems they did not have an efficient method for recycling skeletal materials, so they needed to regrow a hard skeleton after each molt. For a trilobite, the tender time between molts may have lasted days, or even weeks.

Brandt also studied the trilobites' survival rate relative to their segmentation and spinal patterns. She found that trilobites with fewer body segments and fewer complex spines were more successful at molting and survived over longer periods of time. In the end, her work helps to explain not only the decline of the trilobites, but also the ultimate success of other arthropod groups, such as the crustaceans and the insects. All arthropods show varying degrees of body segment fusion, brought about by the process called tagmosis. The general pattern seen across all arthropod groups is that multisegmented ancestors fused their body segments into functional regions, or tagma, and the arthropods with less complex body forms were more successful at molting, and therefore survived better over geological time. Although trilobite diversity rose during the Early Cambrian when all animals were first evolving skeletal forms, trilobites started to decline at the end of the Cambrian, as other arthropods such as crustaceans and multilegged myriapods developed more efficient molting procedures, and while new predatory groups such as fishes, squids, and sea scorpions appeared in the waters.

Now you have a picture of life in the ancient shallow oceans until about 444 million years ago. Arthropods, especially in the various forms of trilobites, skittered about in the sediments, paddled, and floated in the balmy waters. Increasingly diverse faunas of shelled squids, fishes, sea scorpions, and crustaceans continued to feast on the tender molting stages of the trilobites. Then, during the Silurian time, life did something it had never done before in more than 3 billion years of the earth's history. Animals finally set foot on land. Plants sprouted up toward the sun. Terrestrial ecosystems were established. What brought about this remarkable change?

3 Silurian Landfall

All things have beauty, just not all people are able to see it.
ANONYMOUS (fortune cookie wisdom)

If strength and size were everything, then the lion would not fear the scorpion.
(more fortune cookie wisdom)

People of my age vividly remember the events of July 1969 when humans first walked on the moon. We regard them as historically important, and justifiably so. For the first time in nearly four billion years, individuals of a species from earth set their feet in another place entirely, a place so distant and hostile that the challenges of surviving there, even for a short visit, were enormous. Like many of my generation, I remember sitting in front of our grainy black-and-white television, waiting for Neil Armstrong to step off his ladder onto the dusty gray lunar surface. For those of you who are unfamiliar with the term "black-and-white TV," isn't it even more noteworthy that we accomplished this feat at a time when most earthbound viewers didn't have color on their screens? Armstrong's boot prints are so ingrained in our cultural psyche, I'd bet you could sketch their picture. We've all seen them time and again, in books, magazines, posters, and on television.

I propose that there was another day in our history, this one lost in the depths of time, when another set of equally historic footprints were made. But we seldom celebrate or hear about this day in the news. It took place 443 million years ago or more, and like the big bang or a supernova explosion, it was a singular event—the moment when a living organism, an animal, first stepped on the earth.

These earthly footsteps were far more monumental than going to the moon. For the first animals emerging from the oceans and moving onto land, the dry earth was harsh and forbidding. They needed a structural vehicle capable of making the trip: a skeletal system able to sustain the stresses of the terrestrial environment and a locomotion system able to carry them there and back again. They also needed the necessary life-support systems to keep them alive: surface protection from solar radiation as well as extremes of heat and cold, and to prevent water loss, and a respiratory system capable of functioning in a gaseous as well as a liquid environment. Finally, they needed a reason to go there. Life was comfortable enough in the oceans for a long time. What factors motivated animals to move into what seems to have been an impossibly hostile place?

One Small Step for Arthropods

The story of land colonization is usually considered to be the story of the Silurian period, 444 to 419 million years ago. There is evidence that some living things may have been on land before that time. There is even a debate about what it means to be on "land." We'll get back to that point. Suffice it to say that the Silurian is the first age of life where we find abundant fossil evidence of both land animals and land plants. By the end of the period these groups had formed terrestrial ecosystems, at least in marginal, wet marshlands. Nevertheless, these simple ecological systems undoubtedly gave rise to all the later land-based communities of life.

I'm surprised by how often people equate the word "animal" with the word "vertebrates." Recently I came across a science article claiming to be about the "first land animals," but it was about lungfish. Let me make one thing abundantly clear. The arthropods are animals, and they were the first to lift their little legs and step on land, at least by the Early Silurian. The arthropods were best equipped to make the journey. They had the necessary protective gear (external skeletons) and locomotion system (jointed legs) since the Cambrian years. Those lazy, slow-witted, slimy, lumbering lungfish ancestors of ours didn't manage to crawl onto land until sometime during the Devonian—a full forty million years after the arthropods accomplished it. The fact that they were able to do it at all is another contingent event, requiring that some fish just happened to develop enough bony structure in its fins to possibly support its bulky weight on land. It's another coincidence of history, without which none of us terrestrial vertebrates would be here. Again, our mere presence in this story seems nothing less than miraculous.

But as the cartoonist Larry Gonick has adroitly pointed out, we descendents of the lungfishes are the ones who write the history books. And once again it becomes necessary to point out the very subtle human-centrist bias that we have crafted into the history of life, simply by calling the Silurian period the "age of land colonization." We casually and nonchalantly overlook the glaring fact that vertebrates played no role in this drama. For tens of millions of years we continued to paddle around in the oceans, and now we have the unmitigated nerve to imply that the stage was somehow being set for us. Life proceeded quite nicely on land for tens of millions of years without us, and it might easily have done so forever.

Also, by calling the Silurian the time of land colonization, we subtly distract attention from the other major ecosystem: the oceans. We glorify the colonization of land simply because it is a necessary step in the processes leading to the evolution of humans. But the real Silurian news story is the glorious diversity of life in the oceans. The Silurian marks the time of the first coral reefs. These weren't composed of corals like the ones we have today, but of ancient rugose and tabulate coral species that later became extinct. The trilobites didn't go away yet, either; there were still lots of those, along with huge numbers of ammonoid shelled squids and brachiopods and a diversity of fishes. These fishes were mostly jawless, but the Silurian also included the first jawed fishes, the first armor-plated fishes (called "placoderms"— some were up to thirty feet long), and the first freshwater fishes, all of which were also jawless. Before returning to the land, we need to acknowledge that the real pinnacle of biological systems of that time— the peak of Silurian diversity and ecosystem complexity—remained out there in the oceans. We should probably call the Silurian the "age of the first coral reefs."

Although the trilobites were declining in species richness, some of the remaining species were quite common in the Silurian coral reef ecosystems. One particularly abundant trilobite was *Calymene celebra*, which is now celebrated as the state fossil of Wisconsin. During Silurian times, what is now Wisconsin was located south of the equator and entirely covered by shallow seas teeming with trilobites. As a result of these ancient warm seas, the limestone formations of southern

Wisconsin are layered with Silurian trilobites, mollusks, brachiopods, and corals. The Wisconsin trilobite *Calymene* was a bottom-feeder that had the ability to roll into a ball to protect itself from predators, a defensive behavior that may have contributed to its continued success.

Wisconsin isn't the only state to honor a Silurian animal. New York has declared a sea scorpion, *Eurypterus remipes*, as their official fossil. Sea scorpions lived from the Cambrian through the Permian periods, a span of about 250 million years, and although they originated in the oceans, some colonized brackish and freshwater habitats. Sea scorpions are quite notable as probably the largest arthropods that ever lived. Some of the largest species grew to monstrous body lengths of seven to eight feet long. These animals were not true scorpions but more like a predatory version of a modern horseshoe crab. They had a long, sharp, spinelike tail—hence the name "sea scorpion"—but there is no indication that they could sting. They did have large spiny legs for grasping prey, and some had pincerlike claws. So these were probably the first predators that could efficiently feed on the hard-shelled trilobites and brachiopods.

More than 300 sea scorpion species have been discovered from all around the world, but the New York fossils remain particularly important. The first sea scorpion ever discovered was found in 1818 in Silurian rock layers from that state. Around 420 million years ago, the entire area between Poughkeepsie and Buffalo was covered by shallow Silurian seas, and so the rock formations there are so full of their remains that the region is called the "sea scorpion graveyard." Without question, these creatures were among the most spectacular residents of the Silurian coral reefs.

Far less spectacular, but far more abundant and diverse, were the brachiopods, which evolved some thirty thousand species in the ancient oceans. Their common name—lamp shells—comes from the fact that the shells of some brachiopods resemble the shape of an ancient Roman lamp. They also resemble clams, but the resemblance is only superficial, as the two shells of a clam are similar to each other in size, while brachiopods have a smaller top shell and a larger bottom one. Lamp shells peaked in diversity during the Ordovician but retained high species richness over Silurian times. Some of the brachiopods cemented their shells to surfaces to keep them in place, so they were important in building the structure of Silurian reefs. So abundant were the lamp shells in Paleozoic seas that they now are probably the most common fossils in the middle-eastern United States. The very first fossil that I discovered as a child was a brachiopod lamp shell, found protruding from a rock along the banks of the Mississippi River. The state of Kentucky has declared any brachiopod as its state fossil, not bothering to name any particular genus or species; there are just too many of them.

One Giant Leap for Arthropod-Kind

The coral reef ecosystems may have been the biological pinnacle of Silurian times, but since insects are fundamentally terrestrial animals, the story of land colonization must still be told, with a slightly more arthropodan bias. It may have taken tens of millions of years, but eventually species richness on land did outpace that of the oceans; the complexity of our tropical forest ecosystems has vastly outstripped the complexity of our ocean reefs ever since. The pitter-patter of those little arthropod feet echoes loudly across the ages and had profound implications in shaping life's subsequent diversity.

Many biologists have long assumed that plants needed to colonize the land first and to establish ecosystems for animals to occupy. That may not be the case, as some good evidence suggests. Namely, there are trace fossils of arthropod footprints, fossilized tracks, dating to sediments from the Late Ordovician. Even if terrestrial plants were present then, it's clear from the footprints that arthropods were walking out on the open wet soils, quite separate from plants, at the earliest of times on land.

If arthropods were strolling on the beaches more than 443 million years ago, what they were doing there? They may have been avoiding deepwater predators. We must assume that the very first animals to walk on land were arthropods that lived in the shallowest waters, in the intertidal zones. Our longtime companion the moon played a significant role in the evolution of life by creating these pools and the tides that shape them. When the tides ebbed and flowed, any arthropods that could survive on the moist shorelines at low tide would have benefited greatly, simply by avoiding the big predators. As the Silurian progressed, the coral reefs presented an increasingly hostile environment. While the tide moved out, predators that breathed with gills,

FIGURE 3.1. A coiled millipede is a quintessential example of a myriapod: a long, multisegmented arthropod with lots of legs. Creatures somewhat like these were among the first animals to colonize land. (Photo by Kenji Nishida.)

such as sea scorpions, cephalopods, fish, and even large trilobites, swam into the deeper waters. The little arthropods that survived along the shorelines enjoyed a peaceful safe haven, perhaps.

Two groups of arthropods appear to have colonized the shorelines at about the same time: the arachnids and the myriapods. The arachnids were the scorpions and the group from which spiders, mites, and their relatives are descended; the myriapods were long, multisegmented, multilegged creatures, the group from which millipedes, centipedes, and insects evolved. Let's look at each of these animals in turn and consider how and why they might have migrated to the beaches.

Sting Time on the Beach

Among the oldest fossils of terrestrial animals are the first scorpions, dating from the Late Silurian. We may call the Silurian scorpions "terrestrial" because they clearly moved and foraged outside the water along the shorelines, but the prevailing opinion is that they were essentially semiaquatic. They breathed with numerous flat respiratory plates layered like the pages of a book, which are called "book gills." These breathing plates must remain wet to function, so the Silurian scorpions must have moved in and out of the water to keep their gills moist. It was not until much later, in the Devonian, that arachnids developed similar but internalized "book lungs" and became fully terrestrial. Like many modern semiaquatic organisms, Silurian scorpions could probably venture along the shores for extended periods, as long as their gills remained wet.

We can learn a lot about these early land colonists by looking not only at Silurian scorpion fossils but also at modern living scorpions. That's because the living world includes a composite of organisms that evolved at various times in history. Different species evolve at different rates, depending on how they interact with their environments. Well-adapted organisms may not change significantly over long periods of time, so ones that first evolved long ago, like horseshoe crabs and scorpions, are known as "living fossils." That's not to say that scorpions haven't evolved and changed over time. They have. At some point in the Early Silurian there was only 1 scorpion species, and it was aquatic. In the modern world there are more than 1,100 species, and each has unique characteristics. They are all terrestrial, and some have adapted to life in some of the driest conditions, in deserts. But others still require moist living conditions, preferring the earth's tropical rainforests. Still, when you look at a scorpion you are seeing a body form that originated in the Early Silurian with some of the first land colonists.

Scorpions are nocturnal. By day they hide in cracks and crevices, under rocks, and beneath other objects. If the first scorpions were active at night as well, then their pioneering steps onto land were probably taken in the moonlight, to avoid the sun's intense ultraviolet radiation. Remember that the ancient scorpions breathed with book gills and could venture out of the water only for as long as the gills stayed wet.

Scorpions are predatory. They never feed on plants, so these arthropods, at least, could easily have colonized the land well before plants did. Modern scorpions feed extensively on insects, which didn't exist during the Silurian period. What did they eat? If the myriapods occupied the shorelines at the same time, then the scorpions probably ate a lot of them. But if not, there were still plenty of meal choices in the rocky intertidal zone. At low tide, numerous small animals would have been trapped in shallow tidal pools, just as they are today. Soft-bodied animals like annelid worms, small fish, and molting trilobites would have been easy pickings for scorpions, which feed with claw-like chelicerate mouthparts by ripping and tearing their prey to shreds. Scorpions also have large pincerlike claws called pedipalps, capable of manipulating prey and pulling soft tissues from hard shells, and a venomous sting capable of paralyzing small animals. Since brachiopods would have been abundant in the Silurian's intertidal zones, they too were possibly among the early scorpions' prey; if a scorpion could hit a brachiopod's soft parts with its sting, it could then use its pincers to pull the animal's body from its shell.

It is no secret: scorpions suffer from a major public relations problem. We almost universally loathe them, probably for very good reasons. All scorpions possess potent venoms used to paralyze and subdue prey. At the very least their sting is quite painful to humans, while at the very worst it is sometimes deadly. That, coupled with their habit of moving around only in the darkness where we can't see them coming, makes them not very much fun to be around. If you travel in the tropics, you really do need to learn to shake out your shoes in the morning, since scorpions like to hide there.

Some scientists have suggested that humans have an instinctive fear of certain dangerous animals like snakes and spiders. We should probably add scorpions to that list, because the mere sight of one quickly sends many of us into a panic. Maybe we retain some primal, genetically programmed fear of these creatures. Consider the situation for our fishy Silurian ancestors. In the deeper waters, by the coral reefs, they had to contend with the likes of the monstrous eurypterid sea scorpions, and in the balmy shallow waters, they had to contend with the likes of the stinging scorpions. The Silurian was not a very pleasant time for our vertebrate ancestors, and once again, we were lucky to have survived it.

Having said all those nasty things about scorpions, I'm going to give you a reason to like them. The females are really nice mothers. In fact, they may provide the oldest case of parental care. Unlike most female arthropods, which simply lay eggs and let the young fend for themselves, female scorpions carry fertilized eggs inside themselves. The

FIGURE 3.2. A mother scorpion with her babies onboard. (Photo by Piotr Naskrecki.)

eggs take many months to develop, and eventually a female gives live birth to anywhere from six to ninety tiny baby scorpions. Looking like miniature versions of their mother, they crawl onto her back, where they ride around for a week or more. The baby scorpions stay under mom's protection until they have completed their first molt, then they wander off on their own adventures.

Just because they are nice mothers doesn't mean that female scorpions are necessarily nice wives, however. In addition to being dangerous, they tend to be larger than the males, who seem to show an appropriate amount of caution and respect when attempting to mate with them. During their elaborate courtship ritual, a male and female face each other, raise their tails, and move in circles for hours, or even days. Mating eventually occurs indirectly. Male scorpions produce a packet of sperm cells wrapped in a membrane: a spermatophore. When a male deems the time ready, instead of coupling with a female and transferring his sperm cells to her directly, he places his spermatophore on the ground, then attempts to lead her over it. This ancient behavior doesn't sound very efficient, but it seems to work well enough for scorpions, and we see it preserved in some of the most primitive living insects.

The scorpions' reproductive behaviors may provide insight into their Silurian landfall. The spermatophore's membrane helps to slow desiccation, but it needs to remain moist or the sperm cells will dry out and die. Since solar radiation could damage these cells, spermatophore transfer can be more safely done under the cover of darkness. This suggests that scorpions initially colonized shorelines not only to seek food, perhaps, but also to fool around on romantic, moonlit Silurian beaches. The fact that female scorpions retain developing eggs inside their body and give birth to maternally protected live young, however, suggests that the Silurian strands were still dangerous. They may have been comparatively safer than in the deep water, but there were still predators, such as large centipedes, other scorpions, and even larger individuals of the same species, that would have eaten the scorpions' eggs and young.

She's Got Legs . . .

The myriapods, multilegged relatives of the insects, have been present in the background of our story, but we haven't said much about them. You may remember that back in the Early Cambrian oceans, in the Burgess Shale fauna, a few of these leggy creatures scurried along in the bottom sediments. Their body design was very simple: a head up front with one pair of antennae, followed by lots of segments, each with a pair of legs. It's the simplest body plan from which a huge range of arthropod forms can be simply evolved, by a process we've discussed already with the trilobites: tagmosis. By fusing segments, functional body regions can be formed. By modifying legs, an assortment of feeding appendages or mating structures can also be developed. The myriapods, with their versatile body, now become key players in our story, because they are the ancestors from which modern insects evolved.

Three groups of myriapods are worth mentioning here. The first two are quite familiar: the centipedes and the millipedes. The third is a rare tropical group: the symphylans. All three respire tracheally, by transporting air through internal tubes. This suggests that tracheal respiration was an innovation of the first myriapods which adapted to life on land, and that the myriapods passed it along to the insects. Although the centipedes and millipedes tell us a lot about the early colonization of land, they each have specialized in their own ways and evolved into classes distinct from the insects. The tropical symphylans, on the other hand, have a simpler body plan that more closely resembles the anatomy of the ancestors from which insects developed.

The centipedes are perhaps the most familiar myriapod group. There are more than three thousand species, mostly tropical, and they are active mainly at night. Centipedes have thirty or more legs, two per segment, and they really know how to use them: most can run very quickly. Unlike insects, centipedes do not have a waxy cuticle to prevent water loss. They can dry out rather easily, so they tend to stay in moist habitats near soil and avoid direct sunlight. All centipedes are predators, and they capture small animals with their fanglike front legs, which house venom glands. Most feed on other small arthropods, but some large tropical species, up to ten inches long, are capable of killing small vertebrates. Similar to the predatory scorpions, centipedes were certainly capable of surviving in the rocky intertidal zone and feeding on various other small animals long before plants colonized the land.

The millipedes, the leggiest arthropods, are called "diplopods" because they have evolved a unique body type: each segment has two pairs of legs rather than one, and contains two pairs of nerve bundles and heart valves. This shows that their segments formed when two primitive segments, each with one pair of legs, fused together. There are more than seventy-five hundred millipede species, and although they live primarily in the tropics, they can be found all around the world.

Millipedes are a lot nicer than centipedes. If you want a Silurian pet, I'd highly recommend one.[1] They are friendly, they do not have venom or bite humans, and these days it's not too unusual to find some of the giant African species for sale in pet shops. Like the centipedes, however, millipedes prefer to stay out of the sunlight, and so they hide in moss, tunnel in soil or under loose rocks, or live in caves. A few species are known to prey upon other soft-bodied arthropods and worms, but most are scavengers that eat decaying vegetation in addition to fun-

FIGURE 3.3. A white millipede (order Polydesmida) illustrates a unique characteristic of these leggy myriapods: each segment is equipped with four legs. Polydesmids are the largest order of millipedes, with over 2,700 species known. (Photo by Kenji Nishida.)

gal or bacterial accumulations. It appears that the millipedes are yet another arthropod group that was perfectly capable of colonizing the beaches well before land plants evolved; these scavengers would have been able to feed on lots of non-plant-based organic material such as decaying green algae mats, fungi, and bacterial blooms in Silurian microbial soils.

The symphylans have escaped the notice of most people, but they are very important to the insects' story because they most closely resemble the ancestral kind of myriapod from which insects evolved: namely, a short creature with fewer segments than millipedes and centipedes and only two unmodified legs per segment. The symphylans are quite small, only about 2 to 10 millimeters long (less than half an inch). There are about 120 known species, and they mostly inhabit the tropics. Like the millipedes, symphylans live secretively in soil, moss, and decaying vegetation and avoid the sunlight. Modern symphylans feed mainly on decaying vegetation, but like the millipedes, they were capable of living on organic materials in microbial soils before land plants appeared.

These mysterious dwellers in the mosses have a very unusual method of reproduction. Male symphylans produce spermatophores, which they leave on top of long plant stalks. Females need to wander around and find them. Upon discovering a spermatophore, a female symphylan bites it, but instead of digesting it she stores the sperm cells inside her cheeks in special pouches. When she lays an egg, she reaches around and picks it up with her mouthparts, fertilizes it, and proceeds to glue the fertilized egg to a piece of moss.

Green Tide: Plants Colonize the Shorelines

Toward the end of the period, new, taller plants joined the myriapods in transforming the Silurian landscape. Two lines of evidence give us a good idea of what they were like. Preserved fossils from approximately 420-year-old Late Silurian sediments contain the archaic rhyniophyte plants, which are named after an early Devonian genus, *Rhynia*, discovered in Rhynie, Scotland. The oldest one, *Cooksonia*, was the very first vascular plant, and it grew only a few inches tall. Very simple and semiaquatic, the rhyniophytes lived along marginal habitats and had parts that could emerge out of the water. They did not have leaves, flowers, or deep roots, and the more advanced early Devonian species were also relatively short—about 50 or 60 centimeters long (mostly less than 2 feet). The rhyniophytes had creeping stems that ran sideways along the shore, probed tiny root hairs below into the soil, and sent shoots upward from multiple points along their top. Each vertical shoot forked once or twice, forming reproductive structures called "sporangia" at the upper tips. The rhyniophytes' lateral stems allowed them to spread thickly over moist shorelines, since they contained vascular fluid-transporting tissues.

The second line of evidence comes from plant DNA. Molecular studies support the long-held assumption that land plants evolved from photosynthetic green algae and that the nonvascular plants— liverworts and mosses—evolved first, around the Silurian, followed later by primitive vascular plants, such as ferns. Liverworts and mosses require a lot of moisture to survive and decompose rapidly when they die, so they did not fossilize well; however, we can be sure that the Late Silurian shorelines were full of them, as well as the rhy-niophytes and a diversity of soil fungi.[2]

If I haven't said much about plants up to now, it's because the terrestrial arthropods were able to thrive for millions of years before plants arrived and developed the capacity to survive. Arthropods had the initial advantage, because they developed their hard structural parts much earlier. More importantly, being mobile, these animals could pick and choose the time of their land expeditions. Because they're nocturnally active and can easily avoid the sun's harmful rays, the arthropods didn't have to wait for the ozone layer to form before they colonized the land. They just did so under the cover of darkness.

Plants, on the other hand, need sunlight. They didn't have the option of moving ashore at night and hiding by day. This means that plants were not able to survive on land until two things happened: they had to wait for a sufficient ozone layer to develop so they could remain safely exposed all day, and following this they had to develop structural support mechanisms. By the Late Silurian they solved the problem of structural support by evolving the complex molecules lignin and cellulose, and arranging the tough stuff into fluid-transporting bundles. Some scientists have suggested that plants must have colonized land first because they create the oxygen that terrestrial animals require, but the cyanobacteria and green algae had been producing this gas for billions of years before the plants moved inland. Ironically, they—not animals—needed elevated oxygen levels, for the ozone layer's ultraviolet filtering effect and to build lignin and cellulose.

It's fascinating to compare and contrast plants with insects, in terms of how they coped with the difficulties of life on land. Both faced the serious problem of potential water loss, so both evolved cuticles that resist water flow. Since a dense cuticle is impervious to oxygen and carbon dioxide, plants evolved breathing pores, called stomata, which allow gas transfer and can be opened or closed to prevent desiccation. These are directly analogous to insect spiracles. Plants needed to develop a water transport mechanism internally, so they hardened cell walls with water-resistant lignin and built internal pipelines, the tracheids. This is similar to the insects' open circulatory system, a simple arrangement where the internal organs are awash in fluids. Just as insects developed a skeletal system for structural support, plants built woody tissues with lignin and toughened cell walls with cellulose.

But because plants didn't have the option of avoiding sunlight, they evolved complex molecules, the flavonoid compounds, which act as sunscreen and protect living cells from excess ultraviolet radiation. To protect their spores, which were exposed on the plants' highest position, they also evolved another type of sunscreen, sporopollenin.

Some of these plant adaptations influenced insect evolution. Because lignin and cellulose are tough and highly indigestible, they protected early plant stems from potential herbivores. Tens of millions of years elapsed before arthropods figured out ways to consume woody tissues in bulk. The flavonoid sunscreens would have also deterred herbivores. Eventually insects would develop digestive mechanisms to cope with such compounds, and even to build them into their own body defenses, but again that would take tens of millions of years. Only the spores of early plants provided a nutritious, ready food source. The plants defended themselves, however, by placing the spore-forming structures up high, away from millipedes and the like hiding in the soil layer. They also used an herbivore-swamping strategy, producing spores to excess and flooding the environment with more than the plant-feeding arthropods could eat. Millions of years later, in the Devonian period, these nutritious spores may have stimulated the evolution of wings and flight by luring ancient insects high above the ground and giving them a reason to be there.

For a long time, however, the first land animals and plants coexisted peacefully. None of the early terrestrial arthropods were true herbivores. Instead, like scorpions and centipedes, they were predators, or, like millipedes and symphylans, they were scavengers that ate accumulating organic materials in the microbial soils, and maybe some rhyniophyte spores. Modern millipedes and symphylans love to burrow in moss, so the ancient land animals undoubtedly moved into the moss as soon as it arrived. But no evidence suggests that they ate whole plants. My botanist colleagues might get agitated when they hear this, but I like to say that "plants provide a substrate for arthropods." The mosses gave the myriapods a pleasant place to live in and shelter from the sun. The benefit was mutual because in the process of burrowing and feeding, the myriapods loosened and turned the soil, cycled nutrients through it, and conditioned it for the colonizing plants. Contrary to conventional wisdom, the animals may have moved ashore long before the plants, and in order to move inland, the plants needed the animal communities to prepare the soil.

By the Late Silurian, 419 million years ago, the first terrestrial ecosystems had been established. To us they wouldn't have looked like much: the inland areas were still windswept, dry, and barren of life, except for microbes in the soil, while along the shorelines mats of green algae and carpets of mosses and liverworts were studded with rhyniophyte stems rising a few feet up. Nevertheless, while the Silurian rhyniophyte marshlands were not tall by our standards, they provided a virtual miniature jungle for the scorpions, centipedes, millipedes, symphylans, and other arthropod residents. But after nearly 26 million years, the Silurian was coming to an end. The Devonian was approaching, and what changes that would bring. Finally, the plants swept across the lands and rose up tall, and the first forests were established. The planet turned green, and the first insect communities arose. And finally, tens of millions of years after those brave arthropods first stepped on land, our lazy ancestors, the tetrapod lungfishes, hardened their fins, took a deep breath, poked their heads out of the water, and wondered . . . "What's going on up there?"

4 Six Feet under the Moss

We live in a world of insects.

STEPHEN MARSHALL, *Insects, Their Natural History and Diversity*

It's been years since I've lived in Michigan, but I still can't look at the palm of my tetrapodan right hand without fondly remembering the state, and pleasant sunny afternoons spent lazily relaxing on sandy beaches along the shores of the Great Lakes. As any Michigan resident will tell you, your right hand, that modified fin of a Devonian lungfish, provides a convenient map of the lower part of the state. Michigan is, in fact, one of the few states with such a distinctive form that its boundaries can be easily recognized from outer space. It wasn't always that way. The mitten-like form of Michigan's Lower Peninsula was carved out by glaciers over the last 1.6 million years or so. As these ice sheets grew and retreated, they did more than excavate the basins of the Great Lakes. Along the way they gouged deeply into sedimentary layers, including lots of geological formations dating back to Devonian times, 419 to 359 million years ago. These rocks were ripped apart, and smaller pieces were dragged along with the glaciers. The softer stones were eroded and polished by grit and sand in the ice. As the ice melted and the glaciers retreated, pieces were dropped along the way and fell into the chilly waters of the Great Lakes. Over the millennia since then, waves have scoured the ancient rocks along the shorelines with sand. The result is that most of them are polished smooth and rounded and mix pleasantly with the beaches for human feet.

Reflections on Petoskey Stones . . .

My parents moved to northern Lower Michigan in the 1970s, so I've had many opportunities to visit such beaches near Boyne City, Char-

levoix, and Petoskey. They are frequented simply for t'
their water and shorelines and for their soft sand. But t
lar beaches are famous for another reason: they are the
of Petoskey stones, the state stone of Michigan. Petos
actually fossils. They are small bits of ancient coral, b:
fossilized, from great coral reefs that once dominated the shallow ....
that covered Michigan in the Late Devonian, about 360 million years
ago. They are bound to be the favorite Devonian fossil of anyone who
grew up in the state.

The commonness of Petoskey stones gives some sense of how extensive the Devonian coral reef ecosystems must have been. The largest
stone found so far, in the Sleeping Bear Dunes of western Michigan,
weighs more than one ton. Most are much smaller than that. If you
walk in the surf along the beaches near Charlevoix, you can easily spot
them in the water. While the stones are wet, the distinctive form of the
coral shows vividly. The living coral itself has been given the scientific
name *Hexagonaria*, which reflects its six-sided shape. A wet cross section of stone looks like a melting chunk of a bee's honeycomb. If you
pick up a wet Petoskey stone, you'll find that its surface is almost in-
variably smooth. The rocks are soft enough that they are easily eroded
in the sandy surf, and with just some bits of sandpaper and a soft cloth
you can buff them even smoother, revealing the intricate designs of
their hexagonal coral architecture. A polished Petoskey stone feels almost greasy. At times I like to hold one, rub its reflective surface, and
try to conjure images of the lost Devonian period.

The Petoskey corals should remind us of the continuing importance of the Devonian coral reef ecosystems, but discussions of the
period usually conjure other images. Vertebrate paleontologists, when
telling the tale of life, have somewhat egotistically dubbed the Devonian the "age of amphibians." It's the time when our amphibious, four-
finned lungfish ancestors first shambled out of the water. In their defense, I'll note that the vertebrate paleontologists and geologists who
came up with this name were trying to replace a dogmatic creationist
framework with an evolutionary system. Out of necessity, they documented the transition of vertebrate forms over the geological ages, but
in doing so unwittingly set up a new vertebrate-centered historical
system that distracts from many important events in life's history on
earth. Other biologists might tell the story with different emphases.

Botanists will proudly speak of the land plants' expansion, the Devonian origins of the first plants with roots, bark, leaves, and seeds. It was the time of the first forests and the greening of the land, at least along the waterways. A bacteriologist or mycologist would doubtless tell of the proliferation of microbial soils, the development of bacterial and fungal soil processes that made the forests possible. But as an entomologist, I'll speak for the small, hidden creatures—the very first insects—and their surprising roles in these processes.

Despite the proliferation of life on land, which I'll get back to shortly, the real peak of Devonian species richness remained in the oceans among those coral reefs. By the time low forests were first developing along the shorelines, the nearby coral reef ecosystems had developed high levels of diversity and structural complexity. The Devonian reefs were built of tabulate and horn corals, unlike any now living, on which speciose brachiopod lamp shells pasted their forms. During the Silurian and on through the Devonian, the menagerie of crustacean arthropods, ancient relatives of shrimp and lobsters, surpassed the assortment of trilobites. The ecological niches formerly occupied by the trilobites were also being filled by more modern types of arthropods, such as the monstrous eurypterid sea scorpions, which continued to be key Devonian reef predators. The diversity of fishes for the first time surpassed that of trilobites, but the predatory cephalopod "squids" were still more assorted than fishes and continued to exceed them in numbers of species through the entire Paleozoic era.

Although the diversity of trilobites had declined to only about 25 percent of its peak in the Late Cambrian, the population seemed to establish a new equilibrium in the Middle Paleozoic seas. It actually leveled out for tens of millions of years over the Silurian and Devonian times. Then at the end of the Devonian it dropped again dramatically. Although the numbers of trilobite species had dropped enormously since the Cambrian, don't get the idea that they weren't important in the Devonian reef ecosystems. The trilobite species that remained were abundant, and they were well adapted for life in those times. A good example of a successful trilobite from the Devonian reefs is the "frog-eyed" trilobite, *Phacops rana*, a creature so abundant in the shallow seas that once covered eastern North America that it has since been declared the state fossil of Pennsylvania. This particular trilobite had exceptionally large eyes that bugged out on the sides, hence the common name. Presumably those eyes were an adaptation that helped them survive in the predator-rich seas. Like other trilobite survivors, *Phacops rana* had the defensive ability to roll its body into a ball when disturbed—and disturbances would have been increasingly common over the course of the Devonian as fish diversity increased. By the Late Devonian, twelve-foot-long predatory fish like *Hyneria* were common in the waterways. So when those first tetrapodan lungfish poked their eyes out of the water they may have wondered what lay ahead on the beaches, but you can bet that they had a pretty good idea of what lay behind in the deeper waters. Once again, our own relatives were among our worst enemies; predation pressure from larger fish may have driven the lungfish to the shorelines.

Bursting the Balloon: Deflating Our Vertebrate Egos

Recently I came across an issue of *National Geographic* magazine from May 1999. On the cover was a list of articles, including the following: "The Rise of Life on Earth." I thought this title was clear enough, so I turned to page 114, fully expecting an excellent story about the evolution of bacterial life over 3 billion years during the Precambrian. Imagine my surprise when I found the full title: "The Rise of Life on Earth, from Fins to Feet." It was an article about the emergence of the first amphibians, some 365 million years ago. They were the first vertebrates to take a stroll[1] through the mud, so I suppose that we shouldn't be surprised. After all, we vertebrates are telling the stories. But there you have it. Here we are, at the start of the twenty-first century, and we continue to reflect on the ages of Paleozoic time in terms of degree of vertebrate development. Even in our enlightened times, life is sometimes equated with animal life, animals with vertebrates, and earth with dry land.

Sometimes the "age of fishes" is extended from the Ordovician into the Devonian, when fishes diversified greatly. Sometimes the "age of amphibians" is extended from the Devonian, when amphibians first poked out of the water, into the Carboniferous times, when they radiated significantly in the ancient coal swamps. It doesn't matter whether we call the Devonian period the "age of fishes" or the "age of first amphibians." Either way, it's a human-centrist bias that seeks to place our vertebrate ancestors in some kind of elevated position.

Nothing could be further from the truth. During the Devonian, as in all periods since the Cambrian, the diversity of arthropod species greatly outnumbered that of vertebrate species, whether it was in the oceans or in the emerging terrestrial ecosystems. The terrestrial arthropod communities didn't need the amphibians to crawl out of the water. On the contrary, the early amphibians needed the land arthropods for their survival and success. Large amphibians may well have fed on fish, and perhaps on each other, but immature and small amphibians, then as now, would certainly have depended upon small arthropods, such as abundant millipedes in the mossy shorelines, as an important food source.

Because amphibians were capable of feeding on arthropods doesn't mean that they somehow dominated over the arthropod communities. Plenty of already well-established predatory arthropods were perfectly capable of feeding on the amphibians. Small amphibian species, developing young amphibians, and amphibian eggs were all easy targets for aquatic and semiaquatic scorpions. In the mossy shoreline communities there were not only scorpions, but also large venomous centipedes, and now, newly emerging predatory arthropods resembling primitive spiders. The mossy Devonian shorelines were no Garden of Eden for the first amphibians, and once again we should feel very grateful, this time because by some stroke of luck, the Devonian scorpions and centipedes didn't manage to hunt the first amphibians to extinction.

If we could travel back in time and stroll along the Devonian beaches, I don't think I could resist picking up a pocketful of broken corals from the shore. I'm sure that any avid readers of science fiction will tell me this isn't wise, but if I were to see any lungfish or amphibians peeking out of the water, I wouldn't be able to resist tossing some chunks of coral that way, each with a "sploosh," until they went away. Let's take a moment to close our eyes and imagine that scene. Let's toss some broken pieces of coral at those pesky amphibians, and scare all the lungfish back into the water. I don't want them interfering with the rest of this tale.

Into the Woods

The other lead story of the Devonian was the developing land plant communities. Recall that the first land plants did not have extensive root systems and that they needed a very moist environment to re-produce. So, the Devonian "land" plant communities rose up along the shorelines, in marshes and estuaries, along lakes and rivers, and other low areas that retained water, soil fungi, and sediments. Inland and upslope, most of the continental landscapes remained rocky, dry, sun-baked, windswept, and barren of life except in microbial soils.

We can learn a lot about the Devonian plant communities by look-ing at a few kinds of plant fossils from the time. The state fossil of Maine, *Pertica quadrifaria* (an Early Devonian land plant), provides a nice place to start. This is a rare and distinctive state fossil, compared to others that we've discussed so far. About 390 million years ago, a volcano erupted, creating a shower of ash that buried a sizeable ag-gregation of *Pertica* plants. A sample of the *Pertica* "forest" was fos-silized in the layers of volcanic ash and has since been rediscovered in the Trout Valley near Mount Katahdin, Maine. The *Pertica* plants of Maine in several ways resemble upscaled versions of their Late Silurian counterparts, the first vascular plants, *Cooksonia*. Like *Cook-sonia*, the *Pertica* plants lacked leaves, flowers, and deep roots. But they branched more extensively and grew much taller. They were giants of the times, rising to heights of nine feet or more, and unlike mod-ern plants, they didn't have any leaves or flowers, just a lot of forked branches and stems. Photosynthesis took place in the exposed green cells of the upper stems' outer layers, and the tips of the upper stems ended in reproductive sporangia. The main strategy of these early plant communities seems to be branching higher to expose more cells to the sun and placing the reproductive parts higher for better wind dispersal. It makes perfect sense that photosynthesizing organisms would evolve to become taller.

Since plants without anchors would easily be blown over in storms, it should come as no surprise that the next round of plant innovations included the evolution of root systems. The Gilboa Fossil Forest of east-ern New York provides some insights regarding mid-Devonian plants. The Gilboa shorelines comprised a complex plant community, com-plete with low plants covering the ground, small shrubby plants, and small- to medium-sized treelike plants that were fifteen to twenty-five feet tall, with leaves, bark, and roots. Still, the Gilboa vegetation was simple in a couple of ways. The plants had shallow root systems, re-stricting them to growth in marshlands near water. Also, their leaf arrangements were thinly scattered, producing not a dense forest but an open, sunny area. Finally, their sparse leaf structure suggests that production of leaf litter was minimal, so organic material accumulated more gradually than in subsequent forests.

By the Late Devonian, forests of tall trees had developed. Perhaps the best known are those of tropical *Archaeopteris* trees, which were found on all continents, rose to heights of fifty to sixty feet, and had dense thickets of leaves, creating forests of deep shade and thick leaf litter accumulation. They also had deep penetrating roots, so they were able to spread widely over the Devonian tropical lowlands. By the end of the period, a true forest biome had evolved.

On hearing the story I've just described, it's easy to imagine that plants quickly conquered the land. Maybe you visualize mysterious but majestic trees severing the rocks with their roots and triumphantly marching inland. That's not exactly the case. Keep in mind that the transition of plant communities just described occurred over the entire Devonian, a period of more than sixty million years. It's also important to realize that the advancing plant communities didn't just reach out with their roots and crush the rocks. They needed a substrate of microbial soils, which were created not just by plants building organic matter through photosynthesis, but also by the mycorrhizal fungi that grew in close association with vascular land plant roots, thereby increasing the plants' ability to absorb nutrients from the soil, as well as by micro-arthropod conditioning of the soils. These mycorrhizal fungi broke down organic materials and rendered them into forms suitable for plant rootlet absorption. The complex communities of small soil-dwelling arthropods like millipedes fed on the fungi and decaying organic materials, breaking down decaying plant matter, aerating the soils, and moving nutrients by excavating pathways through the soil layers. Other animals, such as soft-bodied annelid worms, moved into the developing substrates, but the arthropods were crucial to the initial processes of creating those soils. In the shade of the first trees, under blankets of the first leaf litter, in complex microbial soils teeming with life, a drama was playing out—and the very first insects were evolving.

Adam Ant?

One of the classic questions in biology is "Why are there no insects in the oceans?" At first glance it does seem puzzling that the most diverse group of land-dominant organisms should be virtually absent there. Many insects have colonized freshwater habitats, and some, such as the brine flies, have extreme degrees of salt tolerance. Only a few insects, such as water striders, have colonized the open seas. But there are no insects in coral reef ecosystems, or other ocean areas, except on the surface or seashore margins. Why? We are so conditioned to thinking about life having evolved in the oceans that we tend to forget that some important animal groups—like the insects—evolved later, on the land. By the time the insects evolved, the niches of marine ecosystems were already filled by other organisms. Various mass extinctions decimated marine communities, but never so completely that the surviving marine organisms couldn't recolonize before the insects could adapt to enter the system. Insects didn't need to move into the already highly competitive ocean environment. They succeeded simply by being the first to colonize each new and unoccupied terrestrial niche.

How did the very first soil-dwelling insects evolve? Full of bacteria, fungi, and arthropod consumers, the Devonian soil microhabitat was not very conducive to fossilization. There are some very important Devonian fossils, which I'll get to in a bit, but for now, let's start by examining living insect species and comparing their anatomy to extinct ones. All insects, whether living or extinct, share a common body form. Through the process of tagmosis, primitive body segments were fused into three distinct body regions, or tagma: the head, the thorax, and the abdomen. An insect head contains the brain to coordinate sensory input, compound eyes for vision as well as simple eyes for detecting light and monitoring changes in day length, a pair of antennae that serve as feelers but allow chemical detection and a sense of smell, and the mouthparts. The insect mouthparts are formed from four primitive leg-bearing segments and are modified enormously across various insect groups for a diversity of feeding functions. The most ancient insects had chewing mandibulate mouthparts, not too different from that of the scavenging myriapods.

The middle part of the insect body, the thorax, is composed of three segments modified for locomotion. All insects have six thoracic legs:

one pair located on each thoracic segment. The wings of modern insects are also located here. Remember that the first insects didn't have wings yet. That innovation came later, in the Carboniferous. The specialization of the insect thorax for locomotion requires that it be packed full of muscles. This is especially true for the winged insects, which require even more muscles for flying. Consequently, there is not so much room left inside that area, so most of the other organ systems are packed into the posterior.

The third insect body region is the posterior multisegmented abdomen. Externally, it doesn't look like much, but internally it is very complex and important. Functional parts of the digestive, excretory, circulatory, respiratory, reproductive, and nervous systems are located there. The heart valves are in the insect's abdomen, and segmentation allows for contractions that increase blood pressure throughout the body. Most of the breathing holes, called spiracles, are located along the sides of the abdomen, whose contractions force air through the tracheal system. This is particularly important when insects molt, as it allows them to pump up their new body form by inflating with air. The systems of the abdomen, including the reproductive organs, are controlled by regional nerve bundles, so even a severed insect abdomen is sometimes able to survive long enough to reproduce. The insects' external genitalia are thought to have evolved from primitive leg parts.

Two Legs Bad, Six Legs Good

Six-legged anatomy evolved during the Devonian, and insects have stuck with that plan ever since: all of the tens of millions of modern insect species are hexapods.[2] Some minor exceptions can be noted. An insect can lose a leg or two in an accident but still walk. Some insects walk on only four legs; the front legs of brush-footed butterflies, for example, have been reduced to nonfunctional, brushlike structures, and they stand and walk on the back four legs. But they spend most of their time flying. Praying mantises' front legs are modified for grasping prey and they walk on the back four, but they are ambush-predators and don't need to walk or run very fast. Other insects, mostly immature forms like maggots, have lost their legs entirely; however, they live in their food, don't need to move much, and eventually develop into adults with six legs. Some adult insects that don't move at all, like scale insects, have lost their legs entirely. Many immature insects, like caterpillars, have evolved additional abdominal "prolegs," which they use to hold tight to their food plants, but they still have the six thoracic legs. Also, no insects walk in a bipedal fashion, as we do, unless you count the brief moment when a jumping insect, like a grasshopper or cricket, stands on two legs as it launches itself in the air. The only bipedal insects are perhaps the fictional ones in animated movies. Cartoonists have humanized these creatures by giving them numerous people traits, bipedal posture being one of the more obvious. I once made a list of the traits of some of the characters from the movie *A Bug's Life* and discovered that some of them have more human qualities than they have insect features.

The fact that there really are no bipedal insects may reveal something important about the reasons for six-legged locomotion. At a fundamental level, it's about balance and stability. I once saw a diaper commercial with a baby falling on his butt, pleasantly cushioned by a fat diaper. The narrator said, "You have to fall about two hundred times before you learn to walk." I'm not sure if there are any data to back that up, but the commercial made a good point: bipedal locomotion is inherently unstable and such balance is not easy to learn. A six-legged insect nymph, however, upon hatching from its egg, is able to walk and run almost immediately. But the six-legged form is not just better for balance; it seems that it may be optimal. Everyone knows how inherently stable a tripod is; insects walk essentially by replacing one tripod with another, moving three legs while keeping the other three planted.[3] The six-legged form is also good for running. Just look at some of the more common groups of primitive insects, like silverfish, bristletails, and cockroaches. They are all very fast on their feet, compared with your average millipede, which needs to coordinate motion across hundreds of legs.

So the process of evolving six-legged bodies seems to be a question of optimizing balance and stability with the potential for rapid motion. Two-legged bipedal locomotion is so unstable and difficult to master that it seems highly improbable and almost pointless. Insects have had 360 million years to experiment with legs, but none has bothered to acquire bipedal form, or four-legged form for that matter, which is stable enough but has less potential for speed. The four-legged tetrapods were all sluggish and slow until they developed warm-blooded metabolism. Six-legged form is sublime. Fifty million insect species can't possibly have it wrong. Eight-legged form isn't so bad either. Just take a look at spiders, which make up thousands of species. But all in all, there doesn't seem to be any real advantage to having eight legs; they're nearly as good as six. And so on with ten legs, or twelve legs, or more. More legs just complicate the issue of coordinating movements, providing no real advantage in speed or stability. Plus, remember the trilobites' downfall. Having more segments and appendages complicates the arthropods' molting process.

It's tempting to think that tagmosis was the driving factor in the evolution of six legs. It seems likely that as with the trilobites, the hexapods would be more successful in part because having fewer segments and appendages would simplify molting. No doubt mastering the molting process and developing a simple body form were key factors in early insects' success, but it can't be as simple as that. Indeed, some myriapods mastered the molting process despite the difficulties of having numerous appendages. Just consider the millipedes' long-term success. They can have hundreds of legs, but they have survived for hundreds of millions of years. Some of the centipedes are fast on their feet as well. So although improving molting efficiency by reducing the number of body segments and legs could be an important factor in the origin of insect form, it can't be the only one. It's also important to consider that the evolution of six-legged form and speed was accompanied by the simultaneous evolution of very small body size. Some of the ancient Silurian and Devonian myriapods were quite large (some millipedes were up to fifty centimeters long). The most ancient hexapods, in contrast, were only a few millimeters long. Over the Devonian, arthropod body sizes underwent very serious downsizing.

The reasons for this trend in reduced arthropod waistlines should be clear enough. The ancient myriapods faced some very serious predators: scorpions, centipedes, and spiders. Natural selection by such macro-arthropod predators would select for smaller arthropods and faster arthropods with fewer legs, since predators tend to hunt preferentially for larger prey, selecting them out of the environment. But being very small has several other advantages. It allowed insects to get deep down into the moss and right into the soil, into moist environments where they could hide and more safely complete their molts. Moreover, small body size also provides a breathing advantage. Smaller animals have more surface area relative to the volume of cells in the body. Therefore the very smallest insects can breathe directly through the cuticle, because they are so small and live in a very moist environment where a thick skeleton is no longer needed. But the ultimate advantage to microscopic body size is that fewer resources are needed for survival. Small animals can grow and reproduce more rapidly than large animals; therefore, they evolve faster, and they can occupy much smaller ecological niches.

There's one other advantage, perhaps, to evolving small body size and migrating deep into insulating blankets of moist soil. The Devonian period, with the evolution of plants with leaves, saw the first accumulations of leaf litter, and consequently the first wildfires. Charcoal deposits from the Devonian document that fires emerged with the first forest communities. With each fire the dry layers of leaves would be burned away, and communities of larger arthropods would recolonize the area. Small arthropods living in the deeper moist soil would be insulated from such local catastrophes.

Springtails Vault to Devonian Superstardom

Probably the most important fossil beds of Devonian-age hexapods are the Rhynie cherts of Aberdeenshire, Scotland. The fossils at Rhynie were formed between 396 and 407 million years ago, when present-day Scotland was a low-lying marsh located in the tropics. Hot spring activity at the Rhynie marsh produced crystal quartz called "chert," which preserved some very small organisms with remarkable clarity; fossil plants from the area, for instance, are preserved in cellular detail. The Rhynie cherts also contain the oldest-known fossils of mycorrhizal fungi, but Rhynie is famous for another reason: among the fossils are the oldest examples of well-preserved hexapods, *Rhyniella praecursor*.[4] *Rhyniella* is also the oldest example of a soil hexapod that continues to thrive in the modern world: the order Collembola, known as springtails.

Apparently the first hexapod group to successfully colonize the finest niches available in microbial soils, springtails get their common name from the fact that they possess an unusual forked taillike structure on the abdomen that allows them to pole vault up to twenty times their body length and spring themselves to safety when disturbed. Spring-

FIGURE 4.1. Springtails (order Collembola) are the most diverse six-legged arthropods inhabiting forest soil and leaf litter. Despite their microscopically small size, they are often numerous and important nutrient recyclers, and sometimes they are quite beautiful. (Photo by Kenji Nishida.)

tails are extremely small, ranging in size from only one to ten millimeters long, and they can be exceedingly numerous: their populations commonly number in the thousands per square meter of soil surface. There are records of as many as a hundred thousand springtails per square meter of soil surface in some remote island shorelines, where there may be fewer predators. Some springtails prey on bacteria, nematodes, tardigrades, rotifers, and protozoa, but most of them scavenge on organic plant materials. Despite their small size, they still account for high levels of arthropod biomass, and because springtails are among the commonest soil micro-arthropods associated with decomposing organic materials, they doubtless contributed in a serious way to the processing of Devonian soils. Their feeding speeds the recycling of nutrients from decaying plants, and their activities improve the physical characteristics of soils, improving nutrient flow and drainage.

Springtails share the same indirect reproduction methods with their myriapod ancestors—male springtails produce spermatophores, which they post on stalks in their moist, mossy environment—and their mating rituals may seem comical to us. Some males spend a lot of time posting stalked spermatophores in a circle around a female, in high hopes that she will pick up at least one, only to see her pole vault away, out of the circle of love. Other male springtails have evolved clasping antennae that allow them to grab onto a female and ride around with her, hopefully until she becomes more receptive. These primitive mating methods would seem to limit the springtails to moist environments, yet modern springtails have evolved to survive under some surprisingly extreme conditions. Some have glycol antifreeze in their blood, and they are the only hexapods known to live along the shorelines of Antarctic islands. On the other extreme, some have evolved to survive in desiccating deserts. They can dry out but rehydrate when it rains.

Because springtails, along with other kinds of ancient wingless hexapods, are microscopically tiny, people rarely notice them. If you want to find them for yourself, there is actually a very simple sampling method, called a Berlese funnel, which takes advantage of their dislike of dry conditions. You can easily and cheaply fashion one of these at home. All you need is a large plastic funnel (like you might use for adding antifreeze to a radiator), some window screen, a jar, some alcohol, and a desk lamp or utility light. Cut a round piece of screen and place it in the bottom of the funnel. Set the funnel on top of a large jar with some alcohol, and add to the funnel a sample of moss, leaf litter, soil, or any such material likely to contain microscopic arthropods. Usually a scoop of any material from the forest floor will work well. Place the light above the wide end of the funnel, close enough to cast some heat and light on the sample, but not close enough to start a fire. As the moss and soil dries, the tiny arthropods will migrate downward to the bottom of the sample. When they get to the screen (if the mesh is not too small) they will fall through, into the alcohol jar. This simple method is one of the easiest ways to see a diversity of springtail species, and may be the only way most of you will see the other kinds of rarer hexapods discussed in the remainder of this chapter.

A Tale of Tails

Other kinds of primitive hexapods survive in the order Diplura (the diplurans). The name literally means "two tails" and refers to the two

FIGURE 4.2. The two-tailed diplurans (order Diplura, family Campodeidae) are reclusive hexapods living in moss and leaf litter, which are seldom seen by humans. (Photo by Kenji Nishida.)

prominent taillike cerci that extend from the end of their abdomens. The diplurans are considered to be closely related to the springtails (they share a unique mouthpart structure—the mandibles are withdrawn into a pouch in the head), but unlike the collembolans, they are scarce and rarely encountered. The only place I have commonly seen living diplurans is in the moist cloud forests of eastern Ecuador, where they, along with numerous springtails, inhabit the dense mosses and soil layers that coat tree trunks and large tree branches, often high above the ground. Moreover, the fossil record of Diplura is very poor, which is hardly surprising since they are soft-bodied and live only in very moist and fungus-ridden microhabitats. Nevertheless, we know quite a bit about their anatomy and biology based on studies of several living species. Less specialized than that of a springtail, their body form is about as simple as it gets for hexapod arthropods: a head with multisegmented antenna and mandibles but no eyes, a thorax with six legs but no trace of wing development, and an elongate multi-segmented abdomen ending with a pair of cerci. Diplurans come in two forms, depending on the shape of the terminal cerci. Some have long multisegmented cerci that look like antennae coming off the tail end, and they use them as back-end "feelers." These species all seem to scavenge organic debris and fungi in soils and mosses. The other kinds of diplurans have cerci whose segments are fused into forceps-like pincers. Many of these species appear to be predatory and some are known to capture small prey, such as springtails or small insects, with their pincer-like "tails." Like other kinds of very primitive hexa-pods, dipluran adults continue to molt their exoskeletons after they reach sexual maturity and adulthood, and some species are known to molt up to thirty times.

Among other primitive six-legged animals thought to have evolved in the Late Devonian are the jumping bristletails, or, simply bristle-tails. As the common name implies, they have long bristly tails, and they can jump by arching their body. Their scientific name, the order Archaeognatha,[5] means "ancient mouth" and refers to their very simple jaw, which is hinged on only one weak point, at the lower head. These jaws are sometimes called milling mandibles, presumably because they can only gently grind food. Bristletail remains have been found in the Devonian Gilboa forest of New York, but several species of these ancient hexapods survive even today in moist forest soils and along shorelines near the oceans. Unlike most modern insects, they are unusual in having abdominal appendages called styli, which may be remnants of ancient leg parts. These living fossils have a few tricks of their own. They use their own feces to glue their bodies to the ground when they molt. This apparently allows them to more successfully pull themselves from their old skeleton during the molting process; how-

FIGURE 4.3. Jumping bristletails (order Archaeognatha) are among the most primitive of living insects. Unlike most insects, they have appendages (styli) on their abdominal segments. (Photo by Kevin Murphy.)

ever, if the glue fails, they can't get out of the old skeleton and they die. It's an indication that early insects needed to evolve efficient molting methods.

Modern entomologists divide the hexapods into two groups depending on whether the mouthparts are withdrawn into a pouch, as in springtails and diplurans, or clearly exposed, as in most other hexapods. By this criterion, the true insects include only the hexapods with exposed manibulate mouthparts: the bristletails, their relatives, and all their descendants. Perhaps the oldest undisputed fossil of a true insect is that of *Rhyniognatha hirsti*, also found in the Rhynie chert formations and aged at 396 to 407 million years old. Although based on little more than a fossil tooth, this organism is particularly intriguing because it has a double-hinged mandible with two hinge joints called condyles, just as in most modern insect species. This type of jaw could belong to a wingless insect like a silverfish or firebrat, placed in the order Zygentoma (formerly Thysanura). However, all flying insect species evolved from ancestors with a double-hinged mandible. This has led some scientists, such as Michael Engel, to speculate that insect flight may have arisen much earlier than previously thought, perhaps as early as the Devonian. Although there are no fossil wings to validate this claim, it is a reasonable possibility. Further, if insects with such refined body plans had indeed emerged by the Late Devonian, then it leads to further speculation that the six-legged body form may have also evolved earlier than usually supposed and that perhaps insects first lived during the Late Silurian. Such thoughts must remain speculative for now, pending the discovery of new fossils that might push the date of the earliest insect back even farther in time. What we do know for sure is that true insects had evolved during the Devonian period, and the terrestrial ecosystems that they emerged in were present at least since the Silurian.

During the Late Devonian, 360 million years ago, complex forest communities of treelike plants had emerged and were widespread across moist tropical lowlands. In the deeply shaded soils below, under thick layers of accumulating leaves, there lived abundant communities of small six-legged scavenging arthropods, including springtails, jumping bristletails, and silverfish. The planet had become buggy. These crawling organisms would seemingly be content to stay in the comfort and safety provided by the soil, the leaf litter, and other mossy, humid habitats. Yet at the end of the Devonian, and onward into Carboniferous times, the world again changed dramatically. At the end of the Devonian the planet cooled somewhat, glaciers formed, sea levels dropped, and the marine reef community experienced mass extinctions. On land, the composition of plant communities drastically altered into the Early Carboniferous period. But the most dramatic change, perhaps, is the first appearance of flying insects during the later part of the Early Carboniferous. By the second half of Carboniferous times, about 320 million years ago, the earth's forests were forever transformed into a fairyland of glimmering and fluttering wings. The next part of the story, the tale of insect flight origin, is one of the most important turning points in the history of life. Where in the world did wings come from?

5 Dancing on Air

In more than 600 million years of animal evolution, there have only been a handful of novel features, such as the wings of insects, that seem in our ignorance not to be mere modifications of something that came before.

DOUGLAS J. FUTUYMA, *Science on Trial*

I have an enduring obsession with wings and flying things. There are few sights more fascinating for me than watching an insect in flight and few tasks more challenging than pursuing one. For as long as I can remember I've been particularly intrigued by the more colorful flying creatures like butterflies and moths, but I don't think that I'm the only one. Judging from the growing popularity of butterfly houses and insect zoos, it appears that most people gain pleasure from the sight of butterflies in flight, even those who detest other sorts of insects. There is a magical, almost fairy-tale quality to them. Is this because we've seen too many Disney movies with delicate little fairies set with insect wings? Or do cartoonists draw fairies with insect wings because dragonflies and butterflies occupy a magical place in our psyche? I doubt anyone would enjoy the fairies so much if we depicted them with bat wings or fish fins.

I've been observing insects for fifty years, but one recent June I witnessed a truly magical event, the likes of which I'd not seen before. My family and I were visiting relatives at their cottage on Canada's Lake St. Clair. We were enjoying a Father's Day barbecue, and by coincidence massive numbers of *Heptagenia* mayflies, which had emerged from the lake overnight, were covering all the trees, the bushes, and the cottages. I'd seen this phenomenon several times before, as a child visiting lakeshores in northern Michigan, but never anything of this magnitude. In the evening, as the sun set orange and fiery over the lake, they rose, shimmering, by the thousands, by the millions, and engaged in a frenzied mating dance. Vast mayfly clouds soared along the shorelines as far as they eye could see, extending ten to fifty feet above the ground, to about fifty feet over the lake. The density of insects was so great that, standing below a swarm and looking up, I could hardly see the sky. Just within my field of view, several million mayflies, mostly males, were dancing in the twilight, slowly rising up and drifting back down in undulating waves. Swarming is a male tactic, a visual display for attracting virgin females from a distance.[1]

Dance 'til You Die

Modern mayflies are truly ephemeral creatures. Since their young nymphs breathe using platelike tracheal gills, they require very clean, cold freshwater for their survival. Because of recent human-induced waterway pollution, massive emergences of many species are increasingly less frequent. Even so, aquatic mayfly nymphs, also called naiads, spend most of their lives hidden in the bottom sediments of freshwater lakes, marshes, and streams, where they feed on algae and organic debris and develop gradually. Once a year, often in synchrony, they emerge from the water, molt, and develop wings from their nymphal wing buds. The adults then have very short lives; many live for less than one day.

Examples of paleopteran insects (a term that means "old wings"), mayflies are among the oldest surviving insects with the most ancient sort of wing design. If you get a chance to see a mayfly, take a close look. You'll observe a relict that developed flight about 330 million years ago in the Carboniferous times. Like most insects, mayflies have four wings: a front and a back pair located on the middle and last segments of the thorax. The front wings are much larger than the back ones, and provide most of the lift for flight. All four wings are simple, however, in that they are capable of moving only up and down; mayflies don't have the ability to flex and twist their wings at the base, as most modern insects are able to do. A mayfly's wing is a delicate membrane overlaying a complex supporting network of many fine veins. Mayflies share this high number of veins with the earliest known flying insects from the later half of the Carboniferous (the Pennsylvanian subperiod). They have another odd feature that is considered to be very primitive: after emerging from water and developing wings,

FIGURE 5.1. A small mayfly, *Baetis magnus* (order Ephemeroptera). Mayflies are considered to be the most primitive living examples of flying insects. (Photo by David Rees.)

they do not fly right away. Instead, mayflies go through a subadult molting stage with functional wings. After one day they molt again, into adults. They are the only living insects that molt after their wings develop, so if you see a winged insect other than a mayfly, you know you are looking at an adult.

When mayflies fly, they do so mainly to reproduce. By synchronizing their swarms, however, they are able not only to attract and meet mates but also to swamp predators. Mayflies are not very strong or adept fliers. They can do little more than flutter their wings and drift and glide in easy patterns. Birds catch them easily, and fish eat their fill as the mayflies land on water. Yet all the predators in the neighborhood can't make a dent in a mayfly swarm. When a virgin female flies into the disco-dancing cloud, she is quickly grabbed by a long-legged eager male, who transfers a spermatophore. Immediately after mating, the female flies back to the lake, then lands and floats on its surface. If she is lucky enough not to be eaten by a fish, the mother mayfly quickly dumps her eggs into the water as she dies. The eggs sink to the bottom and the cycle of death and rebirth is repeated, just as it has been for 320 million years.[2]

When mayflies first fluttered over ancient marshes, the skies were an open frontier: there were no birds or other vertebrates to chase them in the air.[3] But the Carboniferous freshwaters were home to many species of jawless fishes and amphibians. If mayflies returned to streams and ponds to lay eggs, many were probably eaten, so even then natural selection would have favored synchronized emergence as both a predator-swamping and a mate-finding strategy. I'd like to imagine that even Carboniferous mayflies danced in large clouds to reproduce.

Although flight allowed the mayflies to rise in the air, easily find partners, and breed in relative safety, wings also served another valuable function: they helped mayflies to disperse. Nymphs that develop in streams tend to get washed downstream as they feed and develop, and by the time they emerge into adulthood, they are some distance from the egg-laying sites. With their wings, early mayflies could have moved easily upstream, laid eggs, and colonized inland lakes and ponds, which had fewer fish than the deeper waters.

Coal Country Tour

When those first mayflies took to the air some 320 million years ago, they glittered and danced over vast lowland marshes, swamps, and tropical wet forests covered with giant horsetails, club mosses, ferns, now-extinct seed fern trees, cordaites plants, and ancient conifers.[4] These forests were unlike any that exist today, yet we know them well because they left behind abundant fossil plants. The accumulations of these plant materials—the most prolific in the history of life—became most of our natural gas, oil, petroleum, and especially our coal, which are remnants of trees that grew to heights of thirty feet and more in dense wet forest stands. During storms these large trees fell into the moist Carboniferous soils and swamps. Their remains were flooded and buried under sediments, piling up to form deep layers of carbon-rich organic debris. Over time the pressure of geological activity transformed the layers into coal. My university generates electricity from a coal-fired facility, so as I write these words I'm using power produced from some of the last vestiges of Carboniferous plant life. When you drive your car to the supermarket, you might be using some as well.

Why does most of our coal and much of our petroleum date from the Carboniferous period? The orthodox view is simply that the Carboniferous swamps provided optimal conditions for fossil fuel formation. After that time, the world became drier, and conditions were not as favorable. But is that all there is to it? Forests didn't go away after the Carboniferous. If anything, there were even more trees, growing ever more widely at higher elevations, and growing to greater heights over the Mesozoic and Cenozoic. After the Carboniferous, the continents may have been drier inland, but they still had lots of streams, rivers, and coastal wetlands. Trees from dry highlands could easily have washed into rivers and accumulated in lowland marshes. Something other than a change in the weather must have occurred.

Let's consider another question: what happens to a tree when it dies in a modern forest? Birds, such as woodpeckers, chickadees, and nuthatches, excavate cavities in it. When these birds move out, lots of small mammals move in and excavate even larger spaces. Other smaller animals cut tunnels in the dead wood: bees and wasps, for instance, do so not to seek food but purely to build nesting sites. The tree's bark is attacked and colonized by various insects, such as bark beetles, flat-headed wood-borers, and gnawing bark lice. Their tunneling activity loosens the bark and allows fungi to spread under that surface. Fungal growth speeds the tree's decomposition and provides an even more nutritious food source for insects than the wood itself. Larger species of wood-boring beetles lay their eggs on the tree, and their grublike larvae tunnel deep into the heartwood. Meanwhile, down below, bacterial and fungal growth decompose the roots, which small soil arthropods chew on. Eventually the roots are weakened and the tree falls in a storm. This totally changes the local environmental conditions, since much of the tree is now spread across moist soil. It is even more exposed to fungi and small soil arthropods. In tropical areas, the termites move in and tear the wood to pieces. They, along with the wood roaches, are among the few animals capable of digesting wood because of the symbiotic microorganisms living in their guts. In temperate areas as well as the tropics, the carpenter ants will likely move in as well. They do not eat the wood; they simply tunnel and live inside it, but eventually they reduce a large tree to small fragments and wood dust. The pieces are mixed with soil and further reduced by soil fungi and micro-arthropods. In modern forests, there isn't much left that might survive and fossilize into coal or petroleum. Most of the tree is recycled back into the forest's living systems.

In the Early Carboniferous, most of today's macroscopic and microscopic consumers of dead wood had not yet evolved.[5] There were no birds, mammals, bees, wasps, bark beetles, wood-boring beetles, bark lice, termites, or ants. Moreover, during the Devonian and Carboniferous times, plants became tall by producing cellulose and lignin, which are very difficult for animals to digest. None of the earliest insects were able to digest raw wood as well. The giant Carboniferous horsetails, like modern horsetails, toughened their vascular tissues with large amounts of silica, making them virtually indigestible. So the Late Devonian and Carboniferous really were special for their excess production of plant materials, not only because the moist climate and high levels of atmospheric carbon dioxide favored plant growth, but also because the plants were able to produce more biomass than the herbivores could consume, for millions of years. The first important insect wood consumers—the wood roaches—did not appear until the Late Carboniferous. They were followed by the appearance of bark lice and the diversification of wood-boring beetles in the Permian.[6] Over time, increasingly more complex communities of wood consumers evolved, and the global bulk production of plant materials of the Carboniferous has never been repeated.[7]

Maiden Flight

The most notable of the new kinds of vertebrate predators that appeared in the Early Carboniferous are the so-called keyhole amphibians, named after their peculiar keyhole-shaped eye sockets. They were the first four-legged vertebrates to evolve ears. What sounds were these amphibians listening to? I've read that they might have developed sound-producing capabilities and used songs to attract mates or mark territories. That may be the case, but mating is just one aspect of mature vertebrate animal behavior. Ears would have been very handy for more mundane aspects of daily life, like finding arthropod food. The large amphibians may have eaten fish and other amphibians, but the small ones would certainly have eaten a wide variety of arthropods and insects. With ears they could have heard the rustling movements of arthropods moving through the leaf litter, the chewing sounds of millipedes and insects crunching on their meals, or the fluttering motions of newly emerged insects preparing to fly. It's probably no coincidence that vertebrates developed ears at a time when arthropods were starting to make a lot of buzzing and fluttering noises in the forest. The next time you listen to your favorite tunes, perhaps you should take a moment to appreciate the fact that our Carboniferous ancestors liked to eat bugs.

The forests of the Carboniferous years produced more than just our ears and coal supplies. By the Pennsylvanian subperiod (the latter half of the Carboniferous, about 320 million years ago), they housed the very first flying insects. The first step in understanding the origins of insect flight is to consider the ancestral organisms from which flying things evolved. During the Carboniferous there appeared a simple order of insects called Zygentoma, known to us as silverfish and firebrats.[8] Zygentomans are considered the nearest relatives of flying insects, and as their common names suggest, some species have silvery scales, like fish, while others have an affinity for warm places (some are pests in bakeries and live near ovens). Like modern insects, they have a head, a thorax with six legs, and a long abdominal region, and similar to bristletails, the Zygentoma have three long filamentous appendages trailing from the end of their body. They do not have wings or even wing buds, but the upper parts of each thoracic segment are flattened and extend slightly to the sides. Ancient zygentomans probably climbed up on emergent vegetation to molt, and it's been suggested that they could probably jump from their lofty perches, just like modern silverfish.

Zygentoma share some unique features with flying insects, including a jaw that's hinged on two condyles, two cerci, and an infolding of the thorax's side wall—the pleuron—which produces a pleural suture.

This is a crucial trait leading to wing development because it both produces a structurally stronger thoracic area and provides more internal room for muscle attachment. But the similarity to flying insects ends there; zygentoman nymphs resemble tiny versions of an adult, and neither ever shows any trace of wing development. The Carboniferous silverfish, however, did have slightly expanded lobes along the upper sides of their thoracic segments, which some insect paleontologists have interpreted as protowings. They certainly did not have wings, but they may have been capable of brief gliding flight after jumping or falling from perches on high vegetation.[9]

Some fossilized insects from the Carboniferous have flat, platelike outgrowths along the upper sides of the thorax called paranotal lobes. These lobes are larger than those of Zygentoma, but the presumption is that they are modifications of the same structures. In some cases, paranotal lobes are present on the first segment of the thorax, while fully formed wings are present on the middle and last segments. The implication is that wings developed from such lobes, which later disappeared for functional reasons. No modern insect has wings, winglets, or lobes on the first thoracic segment. Instead, their wings are located on the middle or last ones.[10] Winglets disappeared from the front end of the thorax probably because an insect's center of gravity is located more toward its rear, and its first segment is too small to allow extensive musculature.

Because wings appeared at the same time that forests became widespread, some entomologists assume that insects evolved flight by climbing tall plants, jumping off, and gliding with their flat plates—an idea known as the paranotal lobe hypothesis. But why would soil insects, which were perfectly safe and happy in the leaf litter, bother to climb way up on top of plants? Entomologists usually suggest that they may have been feeding on the developing spores, seeds, and tender photosynthetic tissues. This idea troubles me for a couple of reasons. First, modern silverfish don't feed in that way. Second, spores and seeds would fall to the ground anyway, so why bother climbing? I can think of a number of better reasons why insects might have climbed plants, even if they didn't feed there.

Insects are cold-blooded. If nighttime temperatures are chilly, their bodies cool down, and they don't get moving again until they warm up. Modern insects often solve this problem by climbing plants, perching

in the sunshine, and using their wings as solar panels. The larger structural veins of the wings are hollow, so blood flows into them, allowing
heat to transfer back into the body. Even small protowings would have
had the potential to transfer valuable heat, possibly before the panels
could be used for flight. Any insect needs to warm up first before it
can walk, gather food, search for mates, or avoid predators, and whoever starts moving first holds the advantage. This solar panel hypothesis for insect flight is particularly compelling when you consider the
prevailing climate of the Carboniferous. We know that the Devonian
ended with a global cooling period and that glaciers formed over much
of the southern continents during the Carboniferous. Tropical regions
remained moderate and free of ice, but the average climate would have
been much cooler than during the Devonian. There needs to be some
explanation of why soil-dwelling insects would leave the comfort and
safety of the soil layer, and the quest for sunlight and heat might provide a plausible solution.

Or maybe they were looking for mates. Some male insects wait for
females at feeding sites (the dinner and dancing, disco-bar strategy),
while others wait for young virgin females near emergence sites (the
Lolita strategy). When these tactics don't work, lots of males go to a
prominent landmark (the hilltop lover's lane strategy). Many modern
insects move to the highest tree, rock, or hill in the area, so it's possible that Carboniferous insects met mates by climbing to the tops of
the tallest plants.

Once on those plants, small protowings possibly played a role in
courtship and mating. We know that some modern insects produce
wing vibrations and use these sounds as courtship signals and that
others use their wings to actively disperse chemical scents called
pheromones, which signal and attract mates; even a small protowing
might have been used in these ways. Moreover, fossil impressions indicate that some of the Carboniferous insects had distinctive pigmenta-
tion patterns on their wings. In some instances, these patterns included
dorsally visible spots or disruptive banding that may have served to
warn off predators. We can't be sure of these patterns' exact function,
but it's logical to assume that ancient winged insects, like modern ones,
also used them to attract potential partners from a distance.

For whatever reasons ancient insects might have climbed plants,
once they were up there, they needed to get down again, and what

simpler way than to jump? Some tropical frogs climb trees in the evening to feed in the forest canopy then return to the ground at dawn by jumping and gliding. If frogs can safely glide from treetops, then why not insects, which are so small that most could fall some distance without harm? Some scientists have suggested that because the Carboniferous atmosphere was dense with oxygen and carbon dioxide, insects could easily learn to fly. Others have suggested that even the bristletails' tiny thoracic lobes might have provided them with simple gliding flight capability.

On the other hand, some entomologists have suggested that insects didn't need to climb plants to learn how to fly: instead, wings might have evolved from gills and played a role in breathing. This is an old idea that enjoyed popularity more than a century ago. In fact, entomologists developed the paranotal lobe hypothesis to challenge it and explain how wings might have evolved without involving respiration, since modern insect wings do not have any respiratory function. But over recent decades the gill hypothesis for wings has gained some new supporters and maintained opposition. Let's consider its pros and cons. The large, hollow veins of modern insect wings do contain tracheal tubes, so gas as well as blood flows through them. The mayflies, as well as fossil insect wings from the Carboniferous, show us that the most ancient flying insects had enormous numbers of wing veins; presumably they had far more wing tracheae than modern insects. We also know that the two most primitive surviving groups of old-winged insects, the mayflies and the dragonflies, both have aquatic immature stages that breathe with gills, which work by allowing gasses to transfer across a thin moist cuticle. Some soil arthropods, like the springtails, can respire through their cuticle because it too is very thin; perhaps ancient insects might also have breathed across the thin cuticle of their winglets. Because the Carboniferous coal swamps were very moist and the air's oxygen level was exceptionally high, it does seem possible that the ancient insects' protowings served as gills before evolving into wings.

In 1994, Penn State's Jim Marden, along with his colleagues, suggested a novel idea—the surface-skimming hypothesis—based on the observation that some adult stoneflies (Plecoptera) can use their wings to sail across water. They proposed that ancient aquatic insects might have first used protowings or gills for surface skimming, and later evolved the capacity for powered flight by fluttering up from the water's surface. Even so, the gill hypothesis has some serious problems. Most notably, although modern insects may have wing tracheae, none are known to actually breathe with them. Instead, when insects molt to the adult stage, air flows into the tracheae and helps pump up the wings to full size. There are other difficulties as well. While surface skimming has been observed among the stoneflies, it has not been observed among the aquatic mayflies and dragonflies, whose gills are on their abdomen, not on their thorax. Silverfish, the most likely ancestors from which winged insects evolved, dwell in the leaf litter and soil. Also, one of the most successful groups of Carboniferous flying insects, the now-extinct Paleodictyoptera, had terrestrial nymphs.[11]

Personally, I'm content with the paranotal lobe hypothesis, but the controversy may never be resolved to everyone's satisfaction. A gap in the fossil record between the early ground-dwelling hexapods in the Late Devonian and the flying insects of the Late Carboniferous renders wing origins all the more mysterious. On the one hand, we need to consider the possibility that Carboniferous conditions enabled early insect wings to respire, but that they lost this ability as insects developed more modern wings. On the other hand, it is apparent that insect wings might easily have developed for several other good, entirely independent reasons: for gliding, heat transfer, or mating display.

The various hypotheses for wing origin shouldn't distract us from the bigger picture: wings evolved by the end of the Carboniferous' Mississippian subperiod (about 327 million years ago), and once they appeared, they became common fairly suddenly in geological terms. Over the course of several million years, flying insects burst into the air with a diversity that exceeded the Cambrian explosion. They have dominated terrestrial ecosystems ever since, so wings were one of the great insect evolutionary innovations that promoted the insects' disproportionate success in the living world. Keep in mind that they had total command of the air for more than a hundred million years, at least.[12] There were no birds yet, and the best any vertebrate predator could do would be to snap at them from the ground. Occupation of the airways provided the insects with brilliant new opportunities for dispersal, escape, courtship, and exploiting feeding sites. Initially, they didn't have to be very good at flying. For the first plant-feeding insects nibbling on spores at the tops of tall plants, just a short fluttering flight would allow them to easily move from plant to plant. Compared to a crawling insect, which would need to walk all the way back to the ground, find another stem, and climb back up again, a flying insect could save an enormous amount of time and energy. And it would run into fewer predators in the process. Even a gentle flight would allow one to occupy the air and perform a safe mating dance, or move to a new habitat when the swamps dried up or wildfires moved through the area.

Classic Fashions Never Go Out of Style

The most ancient wing style—a flat panel of skeletal material set into a membranous area at the top side of the thorax—was very simple but highly efficient, and some modern insects such as mayflies and dragonflies still sport it. The middle of the wing balances on a fulcrum formed from the pleuron. The wing itself acts as a lever, snapping up or down, because it is stable only in the most raised and the most lowered positions. Its upstroke is accomplished indirectly by muscles that connect the thorax's top and bottom walls but do not attach to the wing. When these dorsoventral flight muscles contract, the thorax is flattened and distorted, causing the wing to snap into an upright position. In the paleopteran insects, the downstroke is accomplished by muscles connected to small plates below the wing, just near the front and back. These are sometimes called direct flight muscles, because they connect directly to the wing. They not only power a downstroke, but by differential contractions they can allow the wing to be tilted at different angles during flight, allowing directed navigation.

Although the mayflies and dragonflies are the only surviving insects with this ancient flight mechanism, when it first evolved it was the latest and greatest innovation in animal locomotion. Without any competition from birds, bats, or flying dinosaurs, the old-wing insects took to the air in prolific numbers, and by the Late Carboniferous, the earth's wet tropical forests were populated by a startling array of flying insect species, almost all of which are now extinct. One of the most notable examples is the order Paleodictyoptera, the old net-winged insects. As the name implies, these insects had a netlike profusion of veins along their old-style wings, which they held out to their sides and were incapable of folding back over their body. But the old net wings conquered the air more profusely than most other insects of the time. Over the Late Carboniferous and continuing through the Permian years there arose a dynasty of paleodictyopteran insects that diversified into at least seventy-one genera, classified into twenty-one families. Some of these species were quite enormous, the largest being about fifty-six centimeters in wingspan (about twenty inches across). They were positively gigantic compared to ancestral flightless insects, clearly indicating that the old net wings had successfully managed to escape the earthbound predators on the forest floor and specialize in tree dwelling, making them some of the first arboreal insects. In the future, most insect species would evolve to live in rain forest canopies.

Part of the old net wings' success stemmed from their invention of a unique mouthpart style. The oldest known true herbivorous species, they were the first flying insects to evolve piercing and sucking mouthparts (beaks) capable of liquid feeding and tapping into the more nutritious plant parts in ways that no other animal group had done before. Species with long beaks and their nymphs could feed on plant tissues and fluids by piercing soft foliage or tapping directly into xylem and phloem. Other paleodictyopteran species with shorter, broader beaks fed by piercing the developing reproductive cones of ancient plants, and sucking out the liquid, spores, and ovules.

We know that the old net-winged insects fed on various plants, including pteridosperms, cordaitealeans, lycopods, and conifers, because we've found fossilized Paleodictyoptera with spores and pollen in their mouthparts and guts. We've also found fossil plant remains that show piercing-and-sucking feeding damage, no doubt paleodictyopteran handiwork. They were the only insects around with the right kind of equipment able to make those marks.[13] The first plant galls, fossils of deformed plant growth of the kind usually caused by arthropod feeding, also date to the Late Carboniferous. We can't be sure what insects made these galls—they could have been formed by mites—but they're possibly the first evidence of insects concealing themselves while feeding inside plants. We also have fossil coprolites, literally fossil insect poop, which contains plant spores.[14]

Many of the paleodictyopteran insects had pigmented patterns on their wings. These species are preserved in detail in some fine sedimentary compression fossils, and although we don't know the colors of the wings, we do know that they often had boldly contrasting

FIGURE 5.2. Fossilized paranotal lobes and patterned front wings of an extinct upper Carboniferous paleopteran insect *Homoioptera gigantea* (order Paleodictyoptera). (Photo by Olivier Bethoux. © MNHN–Olivier Bethoux.)

stripes or spots. What might these markings mean? The most likely explanation is that they provided visual cues for mate recognition, just as in modern dragonflies, grasshoppers, and butterflies. But insects evolve colors for several reasons. It's possible that some of their pigments might have been cryptic. It's difficult to know what sorts of leaf markings or color tones might have been prevalent in the foliage of Carboniferous rain forest canopies, but if paleodictyopteran wings were blends of greens, tans, and yellows, the insects may have been highly camouflaged when resting on tree trunks or in foliage. They may well have escaped entirely from the amphibians on the ground, but they may still have had to contend with the tree-climbing scorpions, spiders, and centipedes. Another possibility is aposematic warning coloration. Many modern insects gain toxic defenses by feeding on toxic plants and develop warning colors in their wings to advertise this to predators. Without any birds, this seems less likely to develop, but paleodictyopterans may still have faced an occasional amphibian on low vegetation, tree trunks, or mossy branches. So there might have been selection pressure to develop bright warning colors, even then.[15] Another possibility is storage excretion metabolism: some insects shunt nitrogen wastes out of their body tissues by packing them into wing pigments (yellow butterflies are the best-known example). Paleodictyopterans might well have had bright yellow or orange colors in their wings, even in the absence of plant toxins or vertebrate predators. The point I wish to make is simply this: the Carboniferous world had pattern, color, and beauty. The orthodox reconstructions of that ancient period overlook this. I've seen several museum panoramas of coal-age swamps, and they always depict a cloudy, drab, green and brown world, usually with a spider, giant dragonfly, and maybe a roach. They never show you the paleodictyopteran insects. I prefer to think that above those swamps flew a shimmering fairyland of multicolored insects, many of which we would certainly consider to be beautiful.

My, What Big Wings You Have

The old net-winged insects may have succeeded magnificently by escaping most of the terrestrial predators in the forest understory, but they still had to contend with airborne insect predators, such as the so-called giant dragonflies, or griffenflies, of the now-extinct insect order Protodonata. Among the most spectacular animals of the Late Carboniferous years, the griffenflies were not really true dragonflies but rather a group that resembled them. Members of one of the griffenfly families—of the tropical family Meganeuridae—are the largest insects that ever lived. During the Permian times, *Meganeuropsis permiana* developed wingspans of seventy-one centimeters (between two and three feet wide), while most other meganeurid species typically had wings four to thirteen inches long. These dragons of the air had sharp, powerful mandibles and spiny front legs for grasping prey. They may not have been really fast or adept fliers, but they were easily able to grab fluttering paleodictyopterans, swarming mayflies, and other flying insects out of the air. They probably also picked off paleodictyopteran nymphs and adults feeding on prominent canopy stems. In the absence of birds, bats, pterosaurs and other flying vertebrates, these Paleozoic giant griffenflies were the dominant predators of the skies, and they were likely the main source of predatory selection pressure shaping the evolution of wing patterns and colors in Carboniferous and Permian insects.

Some researchers have pointed out that the appearance of gigantic flying insects correlates with peaking atmospheric oxygen. Levels had remained around 15 percent from the Cambrian to the Devonian, rising significantly to around 35 percent in the Late Carboniferous. Then they dropped back to about 15 percent by the end of the Permian. Since the times of highest oxygen do correspond to the era of giant flying insects, it is tempting to relate the two. Scientists have suggested that such large insects needed heightened levels to operate their huge flight muscles, and there is some evidence that they had enlarged tracheal systems. Carbon dioxide was also elevated at the start of the Carboniferous but dropped dramatically over the Permian years to nearly modern levels. So it's also been suggested that the air was thicker and more viscous then, making flight somewhat easier.

These physiological arguments may have some problems. They assume that since oxygen reaches insect cells by diffusing from their tracheal network, a respiratory constraint is placed on the insects' upper size limit. But it's important to note that insects can force air through their tracheal system and pump it to cells deep in their body, by contracting their abdominal segments. Also, we don't know what the giant

FIGURE 5.3. A gigantic fossil wing of *Meganeuropsis permiana* (order Protodonata) from lower Permian rocks found in Oklahoma, about 280 million years old. The length of this wing is about thirteen inches. (Photo by Frank Carpenter. Museum of Comparative Zoology, Harvard University. © President and Fellows of Harvard College.)

griffenflies' metabolic requirements really were. They were the top predators, and nothing else was chasing them. Prey like mayflies and paleodictyopterans probably fluttered along slowly, so there is no reason to assume that the giant meganeurids flew very fast. If they had a slow, lazy flight pattern, they may have required less oxygen than some large modern insects, like hawk moths. Another thing to consider is that modern dragonflies are very lightweight, and the giant meganeurids probably were as well. Most of a dragonfly's body is very slender, and it has lots of gas-filled tracheae, which helps a dragonfly to both float on water while laying eggs and to fly more easily. Moreover, there are no modern insects quite so large in wingspan, but there are some massive ones. The heaviest adult insect, the goliath beetle of Africa, can weigh up to a hundred grams. Its thorax is thicker than that of a giant dragonfly, and one of these beetles probably weighs as much or more than a meganeurid air dragon. And goliath beetles can fly just fine, even with the modern atmospheric oxygen level of around 21 percent.

Oxygen might have been a factor spurring the evolution of the air dragons' large size, but it probably wasn't the only one. It's important to remember that insects develop by periodically molting their external skeleton. So for any insect to grow as large as a griffenfly, it would need to go through a series of progressive molts, and during each one it would be extremely vulnerable to predators. The growth of any giant insect therefore requires that the young live in a largely predator-free environment. Where, then, did young giant griffenflies grow up? We suspect that the immature forms were freshwater aquatic nymphs that breathed with gills (like those of modern dragonflies) and were predators who fed in marshes and ponds,[16] where they would have found abundant food in the form of mayfly nymphs or insects that fell on the surface. Like modern dragonfly nymphs, they probably also fed on fish and amphibian eggs, small fish, and amphibian tadpoles. Consider the situation in the Carboniferous years: jawless freshwater fish were spreading to inland lakes and ponds, but the air dragons held the advantage. Because they could fly, the females were able to move inland more easily than fish and occupy temporary ponds and marshes that the fish could not as easily colonize. Griffenflies could also fly inland to unoccupied ponds, ahead of the fish, which, when they finally did arrive, would be attempting to lay eggs in ponds already full of vicious meganeurid nymphs. The large nymphs of many modern dragonflies are able to cover themselves in sediment and debris, and maybe the meganeurids could don similar disguises. Perhaps, as they grew ever larger, they burrowed into soft bottom sediments to complete their molts. If they lived in ponds, the liquid environment would have facilitated their multiple transitions. They would have emerged only to finally molt into fully-winged adults; then they could fly to the comparative safety of the forest canopy, where their main enemies might have been each other.

A New Twist on an Old Style

During the Late Carboniferous, the giant griffenflies started chasing a new kind of insect that was tasty but harder to catch than the old

FIGURE 5.4. Fossil folded front wings with eyespot-like markings of an extinct upper Carboniferous neopteran insect, *Protodiamphipnoa gaudryi*. (Photo by Olivier Bethoux. © MNHN–Olivier Bethoux.)

net-winged ones. Neoptera, or the new-winged insects, were faster flyers, and they had an original trick made possible by tiny articulating skeletal plates, called axillary sclerites, in the membrane near their wing base. These allowed directional wing movements that were never possible among the paleopterans. When neopteran insects landed on a plant and were done flying, they could twist their wings at the base, fold them back over their body, and put them away, making the neopterans much smaller than the older insects, which held their wings constantly outstretched, kitelike. This neat twist would open myriad new possibilities for future insect evolution by allowing front and back wings to specialize separately. It made the future evolution of stoneflies, grasshoppers, bugs, beetles, lacewings, butterflies, bees, flies, and most other modern insects possible.

Some of the neopterans were fast on their feet, too. Quickly after landing, they would deftly fold their wings and run under a leaf or into cracks and crevices, making themselves tough targets for the air dragons. This was such a successful adaptation that before you could say "cockroach," the tropical world was infested with them. Several groups of new-winged insects appeared during the Carboniferous, but the roaches (order Blattaria) were by far and away the most successful of the age. By the Late Carboniferous there were more than eight hundred species, and they made up about 60 percent of the known Carboniferous insects. They were a bit different from modern species: they could actively fly and females had an egg-laying device, an ovipositor, that resembled a tail. We still classify them as roaches, however, and in terms of species diversity, we should probably call the Carboniferous period the "age of roaches."[17]

Roaches have a bad reputation, mostly because of a few bad eggs: a few nasty pest species that overrun our houses and apartments. But please don't base your overall impression on just those few. Modern tropical forests contain thousands of cockroach species, which enjoy a wide variety of habits. Most live on the ground in leaf litter or under logs, but many live on tree trunks or in the forest canopy. Some are blind and live in caves, while others are semiaquatic and live alongside streams or in bromeliad water tanks high in the treetops. Most are nocturnal, but some are active by day, and some even prefer bright sunlight. The night-active species may stir at different times: some in early evening, some around midnight, and others before dawn. Modern roaches are moderately omnivorous, and they play an important role in the decomposition and nutrient cycling of leaf litter and organic material on tropical forest floors. Several living cockroach species are even known to pollinate tropical plants.

Many of the early cockroaches are thought to have preyed on small soft-bodied insects or scavenged the bodies of dead insects, probably of fallen titanic air dragons. Based on the abundance of fossil fecal droppings, many others are known to have been detritivores, and, like modern roaches, are thought to have played a valuable role in the rapid decomposition of leaf litter. The wood roaches evolved a symbiotic relationship with their gut microorganisms and became the first effective macroconsumers of dead wood. The roaches in turn were the most abundant food source for a host of predators, including scorpions, spiders, centipedes, fish, amphibians, reptiles, and the flying air dragons. So with the onset of roaches an important turn occurred in the cycling of organic molecules. More biomass from plant material escaped the geological cycle of sedimentation and rock formation, and was cycled back into the living world by small animals. The great coal age was coming to an end.

In the Late Carboniferous, around 299 million years ago, the sun rose over moist lowland rain forests of giant horsetails, seed ferns, and ancient conifers. As steam rose from the marshes, scurrying roaches folded their wings and nestled into the leaf litter. In the forest canopy, colorful old net wings basked in the early morning sunlight, absorbing heat and fluttering into the air. Giant shimmering gossamer-winged air dragons lifted to the chase. Down below, in the still-shaded shorelines, the amphibians glanced wistfully to the treetops and worried about their next meal. Along the shorelines, also basking in the morning sunlight, was another new animal, one I haven't mentioned yet. During the Carboniferous, the lizard-like vertebrates evolved, and by the end of the period there arose large reptile species that ate meat, mostly from fish and amphibians. But you can bet that small ones liked to eat insects whenever they could get them. You can also bet that whenever there was a massive mayfly emergence, all the amphibians and reptiles would be busy lapping up their fill off the low vegetation.

As hungry reptiles patrolled the margins of the coal swamps, the Carboniferous times were waning. The world was getting drier, and while the coal swamp lands were decreasing, conifers were evolving and transforming the terrestrial landscape. Insects had evolved some of their most important traits—wings and the ability to flex them in a complex way—but now that the Permian times were approaching, some even more startling innovations were developing. During the Permian the giant fin-backed reptiles dominated the shorelines, the flying insects diversified like never before, the most gigantic insects of all time patrolled the airways, and complex metamorphosis became widespread. But the Permian is crucial to the history of life on earth not just because of its evolutionary innovations. The end of the Permian marks the biggest change that life has ever seen: a catastrophic mass extinction greater than any before or since. Understanding that event, and why many insects survived it, may prove to be the most important key in explaining why insects rule the planet.

6 Paleozoic Holocaust

The end-Permian mass extinction had the greatest effect on the history of life of any event since the appearance of complex animals.

DOUGLAS H. ERWIN, "The Mother of Mass Extinctions"

So what did cause the greatest mass extinction in the past 600 million years, and perhaps the greatest in the history of life? The short answer is that we do not know, or at least I do not know . . . . In the wake of the Alvarez-impact hypothesis many of us seem to prefer a single dramatic cause as an explanation for such events. Our knowledge of recorded history provides precious little support for such a view, and I see little reason, a priori to expect such a neat and tidy resolution to this riddle.

DOUGLAS H. ERWIN, *Extinction*

The aim of science is to seek the simplest explanation of complex facts. We are apt to fall into the error of thinking that the facts are simple because simplicity is the goal of our quest. The guiding motto in the life of every natural philosopher should be: Seek simplicity and distrust it.

ALFRED NORTH WHITEHEAD, *The Concept of Nature*

Two film genres from my childhood remind me of Permian times. The first is the giant arthropod thriller. These sorts of movies are not as popular as they once were, but you've probably seen one or more. The classic example is the 1955 thriller *Them*, whose plot involves gigantic ants mutated by atomic radiation. It was the first giant arthropod thriller and, in my opinion, it was one of the best. Many others have appeared over the years. You can take your pick. There was the massive *Tarantula* terrorizing teenagers at drive-in movies,[1] *The Deadly Mantis* with preying mantids big enough to eat army tanks, and *Mothra*, the moth creature whose wing beats created winds strong enough to blow over office buildings in Japan. And of course there was *The Fly*, an all-time favorite about the hapless inventor who has the rotten luck to get insect parts mixed with his own when he enters his innovative transporting device. *The Fly* is perhaps the only example of the giant arthropod thriller that Hollywood valued enough to honor with a modern remake. Finally, there are the more recent but very popular *Alien* movies. In case you were too appalled to notice, the alien had all the key features of an arthropod: a hard external skeleton, segmented body, jointed appendages, and metamorphic development with molting, as well as extendable mouthparts resembling those of an immature dragonfly.

The second genre is the murder mystery, which was very popular when I was a kid. There were the classic novels by Agatha Christie and their movie adaptations, like *Murder on the Orient Express* and *Ten Little Indians*. Who could forget Alfred Hitchcock's *Psycho* and *Rear Window*? The weekly television show *Perry Mason* was a favorite, and judging from the popularity of shows like *CSI: Miami* and *Law and Order*, it seems that America's craving for murder mysteries has not diminished.

Fortunately for us, there are no insects as large as *Them* or *The Deadly Mantis*, and humans thankfully don't have internal parasites as enormous as the Alien. But during the Permian, about 299 million years ago, insects did reach gigantic sizes—not as large as any in these movies, but still impressively bigger than any living today. The fact that these massive insects did exist raises interesting questions about why they lived and why even larger insects never evolved. The Permian also presents a number of unsolved murder mysteries. The colossal insects died during that period, disappearing forever, and the Permian saw the rise and fall of the giant fin-backed reptiles and saw the last of the trilobites. But all these creatures were not alone in their demise. The end of the Permian is marked by the greatest mass extinction in the history of life, when most of the species alive up to that point died. This event is the greatest murder mystery of all time.

Observing the Scene of the Crime and Assessing the Time of Death

The transition into the Permian's early years was gentle. Many of the Carboniferous period's plants and animals continued to thrive, and giant fin-backed reptiles were prominent until the Middle Permian. Among them was the massive *Dimetrodon*, which fed mostly on fish and amphibians, but these beasts probably ate insects too, especially when young. Even a large hundred-pound *Dimetrodon* would have paused for the meal provided by a newly emerged air dragon nymph just crawled up from the water but not yet able to fly. The great fin-backed reptiles dominated the Early Permian shorelines, but in the Middle Permian years they were ecologically displaced by the appearance of new kinds of large vertebrates, including the first warm-blooded animals—the Kazanian protomammals—and the thecodont reptiles, the clan that would later, in the Mesozoic era, lead to the dinosaurs.

Over just a few million years, the protomammals diversified explosively, and what a bunch of noisy brutes they were. They could run, jump, snap, and bite like anybody's business. They scouted out the hills first with slim, fast, dog-sized predators, but before you can say "*Dimetrodon* steaks, a dime a pound," the protomammals diversified into all sorts of terrestrial ecological roles. The dominant large predators, dome-headed Kazanian protomammals, had bony skulls, could roar, and butted heads in fearsome territorial displays. There were also dozens of smaller species that fed on insects; the hot-blooded pursuit of arthropods had begun. So in the upland meadows, while the bugs first tippled on plant saps and the beetles first chewed on logs, it was a brave new world for the vertebrates. Never before had the terrestrial habitats seen such an assortment of these bony creatures. But after ten million years or so, the Kazanians fell by the wayside, and a whole new wave of protomammals terrorized the Permian forests: the dynasty of the Tartarian protomammals. The old dome-heads were replaced by ferocious two-tuskers. Once again a new regime of warm-bloods filled the big animal niches in the Permian meadows and forests, and once again there were lots of small ones that probably ate insects. Then after only a few more million years, the Tartarian protomammals also fell, this time in the days of the end-Permian mass extinction.[2] Overall, about 70 percent of the vertebrate families were lost.

Permian times also saw a startling diversification of insect species, one that surpassed the famed Cambrian explosion in species richness, if not in experimentation with anatomical forms. It is the geological interval with the greatest ordinal-level diversity—there were at least twenty-two insect orders, more than even now—and the old-winged insects reached a peak of diversity unlike any before or since.[3] The Permian years also saw the first of the orthopteroid insects, the ur-crickets and ur-katydids; the first of the hemipteroid insects, the true bugs with sophisticated siphoning mouthparts; and the very first insects with complete metamorphosis, the beetles, lacewings, scorpionflies, and caddisflies. However, the insects experienced more extinction than at any other time in history. Eleven of the Permian's twenty-two orders are now extinct: eight apparently disappeared completely at the end of the period, and three others fell into serious decline and were gone by the Early Triassic.

The plants didn't fare any better. The old Carboniferous and Permian flora faded away. The seed ferns, giant horsetails, and treelike lycopods disappeared, and tall ferns declined greatly in stature and diversity. Plants that had been previously rare moved in to dominate the drier Mesozoic uplands: these included tall conifers, abundant cycads, gingkoes, and new kinds of short ferns in the undergrowth.

The changes in land communities, although dramatic when viewed from our present day, appear to be more gradual when compared to the changes in ocean communities. In fact, the best case for a comparatively sudden Late Permian extinction event involves the ocean reef inhabitants. Our evidence comes from well-preserved Permian and Triassic communities in marine fossil sediments; these fossils show us that more than 90 percent of ocean-dwelling species went extinct, including the last of the trilobites.[4] Some marine groups, notably the trilobites, had been declining long before the end of the period, but the end-Permian events were the final nail in their coffin. It took millions of years, well into the mid-Triassic, for ocean reefs to recover their former levels of species richness and ecological complexity. When they did recover, the new Mesozoic reef communities more closely resembled modern marine ecosystems. Gone forever were trilobites, Petoskey corals, and giant sea scorpions, while the diversity of brachiopod lamp shells and crinoids dropped precipitously.

The Permian extinctions are the ultimate cold case: 252 million years elapsed before anyone noticed that something unusual had happened and started an investigation. From our observation point, the differences between Permian and Triassic communities of life look dramatic and sudden, even if they took millions of years to develop, as was once thought. We are still studying how quickly these changes emerged, but the growing consensus is that the end-Permian extinction was not a singular, momentary event but a prolonged process over a longer period of time, perhaps fifty thousand to a hundred thousand years, that especially affected the oceans. Whether the extinctions happened slowly or quickly, the Late Permian certainly saw stunning upheaval, widespread death, and a remarkable transition to new life. The extinctions and changes in plant and animal communities from the Permian into the Triassic years are so extraordinary that geologists naming the earth's layers decided to draw a special line at this point in time and divide the geological periods into eras of life. All the periods discussed so far—the Cambrian, Ordovician, Silurian, Devonian, Carboniferous, and Permian—are combined into a single time known as the Paleozoic era, the "age of old life." The second era, the Mesozoic, encompasses the Triassic, Jurassic, and Cretaceous periods, the time of the dinosaurs.

The Suspects

Here's a brief list of some of the various hypotheses for the Permian extinctions. New kinds of animals evolved, which displaced older animals that could not compete as effectively for resources. New plants arose, which were better adapted for drier climates. Plant-feeding insects went extinct along with their Paleozoic hosts, and they could not adapt to feeding on new plants. A stupendous event, perhaps massive volcanic eruptions or an asteroid collision with the earth, killed oceanic plankton and catastrophically eliminated the base of the marine food web.[5] Some have suggested that both widespread volcanic eruptions and a massive asteroid impact happened at the same time. Others implicate the changing continents. Plate tectonics altered the continents' positions, changing sea levels, reducing shoreline habitats, and creating drier climates inland. Continental drift combined land areas, bringing together communities of organisms that could not coexist. Then there are the glaciers. Some suggest that expanding ice sheets caused turbulent mixing of ocean waters, which brought toxins to the surface from the depths. Others implicate changing atmospheric gases, either increasing carbon dioxide or decreasing oxygen levels or both. It has been suggested that the Triassic's characteristic red rocks indicate an oxygen-depleted atmosphere. Peter Ward, in his 2004 book *Gorgon*, argues that the massive gorgon protomammals of South Africa died of asphyxiation, while the dinosaur lineages survived by virtue of their more competent lungs. Declining oxygen levels are claimed to have thinned the air, making it more difficult for gigantic insects to fly, and perhaps impossible for them to function at high metabolic rates.

Smithsonian paleontologist Douglas Erwin has written two books and several articles on the subject of the end-Permian extinction. He points out that there are people who think they know what caused the extinctions and that because some of the proposed answers are "mutually contradictory," they can't all be correct. Describing it as "a tangled web rather than a single mechanism," Erwin proposes a *Murder on the Orient Express*-type hypothesis, which suggests that there is not one simple answer; instead, the best solution might be a combination of several of the proposed causes. He notes that the end-Permian events seem to have taken place in three parts. The first involved the widespread drying of shallow ocean basins along with wide swings in climate conditions. The second phase involved chemical changes in the oceans, as well as widespread volcanic eruptions that contributed to a rapid increase in atmospheric carbon dioxide levels. This accelerated global warming and a sudden drop in oceanic oxygen levels—a condition known as anoxia—which had disastrous consequences for ocean life. The third part involved widespread extinction in coastal and near-shore terrestrial communities.

In his 2006 book, *Extinction*, Erwin bluntly says, "I do not know." That's a pretty amazing statement coming from the one scientist who has studied the end-Permian extinctions more thoroughly than any other living person, and it probably serves to illustrate the problem's continuing complexity. But fortunately he goes on to give his best educated guess at the possible solution, echoing his previous thoughts: the Siberian "flood basalt" volcanoes may have triggered the mass extinction by releasing massive amounts of sulfuric aerosols and carbon dioxide into the atmosphere, which could have triggered global warming. Global warming would have caused anoxia and the rapid mass extinction of ocean life. This scenario nicely explains the rapid catastrophic loss of life in the salt-water oceans, but the events on land with terrestrial insects, and in fresh water with aquatic insects, remain more mysterious.

FIGURE 6.1. A beautifully patterned fossil of *Dunbaria fasciipennis* (order Paleo-dictyoptera) from Permian rocks of Kansas. This paleopteran (old-winged) insect order was one of the casualties of the Permian extinctions. (Photo by Frank Carpenter. Museum of Comparative Zoology, Harvard University. © President and Fellows of Harvard College.)

Interviewing the Survivors

One of the prominent features of the Late Permian insect extinction is that the old-winged paleopterans were severely affected. Most did not survive. Four orders went entirely extinct: Paleodictyoptera, Megasecoptera, Diaphanopterodea, and Dicliptera. All of these insects had a sucking beak and immature nymphs that were terrestrial. By contrast, the groups of old-wings that did survive the Permian extinction all had freshwater aquatic nymphs with gills that developed in ponds, streams, and marshes. The order Ephemeroptera, the mayflies, were quite diverse in the Permian, with at least five families, and they survived to the present day. There were also at least six families of Permian damselflies (order Odonata), delicate creatures related to modern dragonflies, and they became highly diversified during the Mesozoic dinosaur days. Then there is the order Protodonata, the giant air dragons. The most gigantic species went extinct during the mid-Permian, but other smaller species survived well into the Mesozoic, flying over the first dinosaurs.

These details suggest that although the end-Permian extinction massacred life in the oceans, insects living in freshwater ponds and streams seemed to find sanctuary. This is further supported by the observation that two other more specialized groups of freshwater aquatic insects also survived: the stoneflies (order Plecoptera) and the caddisflies (order Trichoptera). Therefore, any models of the Permian extinction need to account for events that would have affected saltwater but not freshwater systems. Scenarios involving dropping sea levels, loss of coastal marine habitats, sedimentation, rapid turnover of deep-sea toxins, and depletion of oceanic oxygen levels all make sense. On the other hand, scenarios involving comet or asteroid impacts don't seem to make much sense, since these impacts should have severely affected freshwater habitats. This is particularly true because the aquatic young of mayflies, stoneflies, and caddisflies are notoriously sensitive to environmental change. Their survival clearly indicates that some fresh water habitats were scarcely affected.

If life was somewhat easier for the aquatic insects during the Permian, then a corollary might be that life was more difficult out of the water, in the terrestrial forest biome. The old-wings faced increasing and ever more novel insect competitors. Most notably, the new-winged insects proliferated extensively. The cockroaches, order Blattaria, we know already from the Carboniferous years. They declined somewhat in the Permian, but are still among the more common fossils from the period. The order Protorthoptera, a motley assortment of cricketlike insects that had chewing mouthparts and the new-winged design, were enormously successful, comprising at least forty families, and over the Middle to Late Permian, they had the greatest species diversity and abundance. The order Orthoptera, a brigade of crunching and munching insects, with strong mandibulate jaws, that includes the ancestors of modern crickets and katydids, reached a new frenzy of activity during the Permian years, ascending into the ranks of the most diverse insect orders.

If you've ever had a roach or cricket in your house, you know that they are fairly aggressive and remarkably omnivorous. They will chew on all sorts of organic materials. We tend to think of grasshoppers and katydids as being vegetarians, but many are also actively carnivorous when they can catch other insects, and others are notoriously cannibalistic. So, you can bet that the Permian protocrickets, roaches, and ur-katydids all munched on old-winged insect nymphs whenever they could catch them. And catching them would have been easy enough. Slow-moving paleodictyopteran nymphs, which would climb up on conspicuous plants and insert their beaks into mature sporangia to suck out nutritious spores, would have been easy targets for the predatory orthopteroid insects. The orthopteroids would have also competed for the same foods more effectively. While the old-wing nymphs could only suck spores from maturing sporangia, the orthopteroids could chew up and consume entire sporangia even before they had matured enough for the old-wing nymphs to feed on them. Finally, the nymphs that did mature to adulthood would have had a real disadvantage in an increasingly new-wing–dominated world. Because the old-wing insects could not fold their wings and hide in small spaces or fly as fast as the new-wings, they would have suffered greater mortality simply because they provided a larger, slower target for predators of all kinds.

One other surprising Permian success story is the icebugs. Also known as rock crawlers, these creatures comprise a small extant order called Grylloblattodea, which means "cricket-roach" and refers to the similarities they share with their cousins, the crickets and the roaches. But the icebugs established their own, very unique lifestyle. Sometime during the Permian years, the ancestors of the rock crawlers presumably moved up along the streambeds to higher and higher altitudes, and eventually adapted to life at the uppermost elevations near permanent ice and snow fields. In alpine settings wings become a detriment—flat parts that can gather heat also can lose heat—so the ice bugs, along with other high-elevation insects, adapted by evolving a wingless body form.[6] Secondarily wingless insects are commonplace today—think lice, fleas, and worker ants—but the icebugs were among the first insects to give up their wings. To keep warm, they developed dark pigments for absorbing solar radiation, and useful behaviors like hiding under flat rocks that face the sun. They are also slow to develop.

FIGURE 6.2. This very cryptic katydid from Ecuador (family Tettigoniidae) is a modern example of the highly successful insect order Orthoptera, which began their rise to prominence in the Permian. (Photo by Angela Ochsner.)

An icebug might take several years to mature to adulthood, because it's cooled down and inactive far more often than it's active. When the alpine weather is mild and the icebugs do become active, they emerge from their warm hiding places and scavenge for food over the surface of glaciers and snow fields. You might suppose that there is not much to eat on top of glaciers, but flying insects don't always end up where they want to be. Many are caught in storms, and clouds of insects are often blown great distances, some to high elevations, where they freeze and fall to the ground. Later, in mild weather, grylloblattids scavenge their flash-frozen carcasses.

They haven't diversified much over the years, but the icebugs show us that you don't need to be highly diversified to survive global catastrophes. From Russian Permian fossils we know that the stem group of grylloblattids evolved more than 252 million years ago, and that ancient icebugs somehow survived the end-Permian and end-Cretaceous extinctions without mishap. They have been with us ever since.

Life Sucks: The Liquid Feeders Tap into Success

One of the greatest Permian innovations was the homopteran piercing-sucking mouthpart design. This adaptation consisted of hollow needlelike feeding tubes called stylets, and it made the Homoptera[7]—the ancestors of modern cicadas, planthoppers, leafhoppers, froghoppers, treehoppers, aphids, scale insects, and their relatives—one of the most successful clans of insect plant feeders. Their very refined stylets allowed the homopterans to drill deep into plant tissues, inject salivary enzymes through one set of tubes in order to predigest tissues and fluids, and extract liquids through the other set, often directly from the plant's vascular transport system: the phloem. Just when the Permian climate was getting drier, these insects invented a way to feed on highly nutritious liquid food. It's also an innovative way to avoid plant defenses. Homopterans don't have to worry about highly indigestible molecules like lignin and cellulose; they just avoid eating them entirely. At the same time these insects are able to avoid the flavonoids and other secondary defensive chemicals that may accumulate in leafy photosynthetic tissues.

The homopteran insects might seem to have invented the perfect feeding system, but there was one problem: too much of a good thing. Liquid food is saturated with water. Dissolved food molecules needed to be concentrated and its excess water dealt with. The homopterans solved this problem too in an even more fascinating way: they evolved a filter chamber in their digestive tract that concentrates food and rapidly shunts away excess water. The filter chamber works like this: a homopteran's digestive tract is very long, and its back part loops around over its front section. A membrane wraps around the place where the back portion overlaps the front, and at this point, two things happen: excess liquid water is transferred directly into the hind intestine for rapid excretion, and food molecules are concentrated for the ride into the middle gut, where they are absorbed into the body. The digestive mechanism is not perfect, the result being that homopteran insects are constantly excreting massive amounts of water with traces of sugars. If, in the summertime, you have ever parked your car under a tree infested with aphids, you may have experienced this phenomenon. The sticky drops on your windshield are aphid poop. In dry climates this stuff tends to crystalize into solid chunks on tree branches, which later fall to the ground; this is the origin of the term "manna from heaven."

The homopterans diversified greatly over the Permian years, survived the end-Permian without any apparent mishap, and evolved rapidly over the Mesozoic era. All this occurred before flowering plants with nectar and sweet fruits evolved, so if the earliest dinosaurs ever craved a sugary snack, homopteran insects' manna would have been the only sweet food around. The homopterans' proliferation once again suggests that the end-Permian extinction event was not likely a singular catastrophe. Instead, their success is consistent with the global climate change hypothesis. Maybe the climate on land was getting hotter and drier, but homopterans succeeded by developing a liquid diet and modulating fluid uptake. Maybe plant communities were changing in ways that the ancient insects could not deal with, but homopterans succeeded by tapping directly into nutritious, undefended juices deep in the plants' vascular transport tubes.

The order Thysanoptera, known as thrips, also evolved a liquid feeding strategy at about the same time. Their scientific name means "fringed wings" and refers to the thrips' very narrow wings, which are fringed with long hairs. However, this feature is not unique to them. Feather-winged beetles and the microscopic wasps known as fairyflies also have fringed wings. The really unusual thing about thrips is their asymmetrical mouth: the mandible is missing on the right side but is present on the left, which means that they have one left tooth, but nothing to oppose it. Thrips use that one tooth very creatively: they twist their head when they feed and use their tooth to scratch the cellular surfaces of tender plant parts. Once the plant is damaged, nourishing fluids ooze from its surface. Thrips then use their remaining mouthparts, formed into a short fluid-sucking beak, to tipple on the liquids. So the thrips evolved a way to tap into tasty plant juices without tapping into vascular tubes, thus allowing them to avoid the problem of intense fluid pressure faced by the homopteran insects. But their feeding style required them to evolve microscopically small body sizes; most species are only a few millimeters long or less. This combination of fluid feeding and minute body size turned out to be a real winner for the thrips, which apparently cruised through the end-

Permian mass extinction without serious problems and have achieved moderate species diversity over the past 250 million years. Currently, there are at least 5,500 living species.

The Transformers: The Rise of Complete Metamorphosis

The proliferation of the orthopteroid crunching and munching proto-cricket brigades, along with the refinement of fluid feeding by the homopteran bugs, might seem to be the major insect accomplishments of the Permian, but another insect feat ultimately makes these things pale in comparison: the diversification of complex metamorphosis, which occurs among insects with larval stages. This innovation may arguably be the single most important factor in the insects' long-term success, as more than 90 percent of modern species belong to groups with complex, or holometabolous, metamorphosis.[8]

At a fundamental level, the success of complex metamorphosis boils down to wing anatomy. The new-winged insects, with their flexible wings, had a great advantage over the old-wings. But they still had a disadvantage: young nymphs developed wings from buds exposed on the sides of their thorax, and as they gradually metamorphosed, they risked damaging their growing wings. The holometabolous insects solved this problem by internalizing the development of their wings, which, along with other adult features, stem from cell clusters called the imaginal discs. This allowed young holometabolans to lead active and even aggressive lives, and avoid damaging their growing wing parts. But internal wing development had other important implications. Holometabolan larvae could burrow into plant tissues, fungi, dead animals, or any other substrate, also without damaging their developing wings. They became highly efficient feeding machines, to the point where many adults did not need to eat (they simply relied on stored food reserves from the larval stage); such feeding specialization allowed larvae to eat entirely different foods, effectively taking adults out of competition with their own young for food or habitat space. Adults in turn became more highly specialized for the mature tasks of courtship, mating, egg laying, and dispersal.

To facilitate the change from feeding larva to reproductive adult, a transformational stage evolved: the pupa. Many people think that the pupa is a resting stage, but this is far from the truth. It may allow insects to hibernate over cold winter months or during prolonged dry seasons, but its purpose is far more important. Inside the pupa, remarkable cellular changes take place. Muscle systems are restructured and the massive thoracic muscles used for flight are constructed; the wings, reproductive organs, and adult sensory systems are built. The digestive system may be extensively modified and rearranged, especially when larvae and adults eat very different food. A good example of this is the transformation from a caterpillar, which feeds on solid plant materials, to an adult butterfly that feeds on liquid plant nectar. So the advantages of complex metamorphosis are many: it allows for the safe internal development of delicate wings, new feeding possibilities, the specialization and separation of immature and adult behaviors, and the development of diverse resting stages for escaping difficult environmental conditions.

Who were the Permian Holometabola, and what can they tell us about the period's events? Some were obscure, such as the extinct insect order Miomoptera, known from fossil wing fragments so difficult to interpret that some paleontologists do not even agree that they belong to the Holometabola. However, many others should seem familiar because they survived the end-Permian extinction and continue to thrive today. They include the scorpionflies, lacewings, beetles, flies, moths, and caddisflies.

Killers with Long Faces and a Lot of Nerve

The scorpionflies, insect order Mecoptera, are not at all closely related to true scorpions, and they do not sting. Their name comes from the fact that the males of certain species have large bulbous genitalia that resemble a scorpion's stinger. Their greatly elongated lower head, which gives them a horselike appearance, distinguishes them from other insects. Most scorpionflies actively prey on other small insects or scavenge dead insect bodies. The tip of their narrow snout has well-developed mouthparts, which allow them to reach into narrow spaces and chew on small prey. During the Permian years, scorpionflies were the most abundant and diverse insects with complete metamorphosis, evolving rapidly in the Early Permian,[9] and by the Late Permian developing into eleven scorpionfly families, almost double the number that exists today. In fact, the diversity of Permian scorpionflies was greater than at any time since then,[10] and by the end of the period it must have greatly influenced other insect populations in the forest biome, since scorpionflies also commonly preyed on slow-moving insects and insect eggs.

The lacewings and their relatives, order Neuroptera, also appeared in the Early Permian, and by the Late Permian this voracious clan comprised at least six families. The scientific name means "nerve-winged" insects and refers to the abundance of very fine wing veins; hence their common name. Neuroptera actively prey on other small insects. The adults have chewing mouthparts, but the larvae are able to pierce insects with their sharp sicklelike jaws and suck blood though narrow channels in their mandibles. Most neuropterans are terrestrial, but their novel fluid-feeding ability allowed some of them to invade freshwater habitats. Lacewing larvae also evolved another curious and useful trick. All insects have excretory organs called Malpighian tubules that extract nitrogenous wastes from the blood and dump them into the hind intestine for removal, but the neuropteran larvae evolved the capacity to convert their waste products into a useful product: silk, from which they spun protective cocoons for the pupal stage. However, to get the silk out of their bodies, they need to excrete it. Neuropteran larvae are the only animals that spin silk from their anus.

Silk Spinners, Architects, and Geologists

At least one other new Permian insect order could spin silk: the Trichoptera, known as the caddisflies. The name "caddisfly" is thought to derive from the old English "caddice men," vendors of material who pinned cloth samples to their jackets (as a caddisfly larva glues various materials to its portable case). Their larvae spin silk in a more familiar fashion: at the head end, from modified salivary glands located near their mouthparts. Caddisflies used their silk to colonize a new habitat: freshwater streams with rapidly moving water, an excellent place to find food particles washed downstream or other small insects and aquatic animals dislodged by currents. Plus, lots of small tasty mayflies and stoneflies were in the fast-moving streams, eating the algal mats on stones.

Caddisfly larvae used their silk not just as anchors, or safety lines, when moving downstream to capture prey; some learned to weave aquatic nets in the currents to collect organic debris. Others learned to gather small bits of rock, sand, wood, and other materials and weave them onto stones, into protective tents. Eventually some figured out how to make those shelters portable by tying silk and collecting items into movable houses. Observing the portable cases of modern caddisflies (they are among nature's most adept architects) may tell us a lot about the behavior of ancient species. The modern cases are really impressive, coming in all manner of sizes and shapes. Some are long, others are short. Some are round, while others are square or spiral. Each species builds its own unique style of case and collects its own preferred set of building materials: sand, pebbles, large stones, small sticks, chunks of wood, pieces of shell, or pieces of leaves. Some preferentially collect heavy particles, however, and one is known to accumulate and weave gold grains into its cases, making caddisflies the first geologists as well as the first architects.

Aside from their obvious protective function, the cases serve other uses. Some caddisflies tie large ballast stones to their cases, allowing them to move along the bottom in fast currents without washing away. Most build a case with a hole at each end, which allows waste to be ejected from it and water to flow through it. Many have evolved the capacity to ventilate their tracheal gills by actively pumping water through the portable case, thereby increasing oxygen flow over their gills. This has allowed caddisflies to successfully radiate into slow-moving or still waters with much lower oxygen content.

In their classic paper, "Ecological Diversity in Trichoptera," aquatic entomologists Rosemary Mackay and Glenn Wiggins observed that in modern aquatic insect communities, caddisfly species and genera greatly outnumber that of the mayflies, dragonflies, or stoneflies. They wondered why this should be so, and they came to a perceptive and surprisingly simple conclusion, neatly summarizing 250 million years of aquatic insect evolution with this simple statement: "We view much of trichopteran diversity as an expression of ecological opportunities made possible by the secretion of silk."[11] That wonderfully versatile substance allowed caddisflies to divide the aquatic habitat into hundreds of microhabitats inaccessible to other insects without silk. Even though many of the mayflies, damselflies, and stoneflies colonized the waters millions of years earlier, caddisflies were able to spin and weave their way to new lifestyles impossible for the more ancient aquatic insects.

The presence of caddisflies during the Permian suggests that primitive moths (order Lepidoptera) must also have been around, even though they do not appear in the fossil record until the Jurassic period, about fifty million years later. A lot of anatomical and behavioral evidence suggests that the Lepidoptera and Trichoptera are closely related to each other: they are what we call sister groups, which by definition originate at the same time because they share a common ancestor.[12] So this is one of the better documented cases of a major gap in the insect fossil record. We know that moths—or at least protomoths—must have existed at least since the Permian, but clearly they did not fossilize well for another hundred million years. If the most primitive surviving Lepidoptera are any indication, there are obvious reasons for the gap. They are microscopically small species that mine and feed in plant tissue; the most archaic group, the mandibulate moths (family Micropterigidae) feed on fern tissue in extremely moist, nearly semi-aquatic environments. Because microscopic soft-bodied insects living in moist, warm forests decompose rapidly when they die, the earliest moths did not fossilize much, if at all.

The insect order Diptera, the true flies, also originated in the Late Permian years, and although they were not very common then, they somehow managed to survive the Permian extinction and live on to become some of the most common and diverse insects in the modern world. Like caddisflies, ancient nematoceran flies had aquatic larvae that lived in cool, fresh, fast-moving water. These larvae developed various suction-cup holdfast structures for clinging to rocks in fast currents, where they fed on algae and organic debris. To this day, some of the more primitive aquatic fly larvae in existence can spin silk, which they use to anchor their bodies in a current or move safely downstream.

Why were streams so popular among Permian insects? During the period, the southern supercontinent, Gondwana, experienced extensive glaciations. Continental areas were colliding and inland areas were being raised up to greater heights. In areas were glaciers met temperate and tropical climates, melting ice and snow from upper elevations created several cascading waters, which offered a rich new frontier of streambed nutrients for insects that could adapt to the swiftly moving currents and eddies. Mayflies and stoneflies were the first colonists to follow the streams up to higher and higher elevations. Soon they were followed by species of caddisflies, nematoceran Diptera, and aquatic predatory Neuroptera. Whether the Permian was a grand disaster or a time of plenty just depends on your point of view. For insects that were able to find and colonize new niches it was a time of grand success. The aquatic mayflies, stoneflies, caddisflies, and nematoceran flies all successfully survived the Permian and have radiated extensively since then.

Meet the Beetles and Other Bugs That Bite Their Bark

From the first humble beetle (order Coleoptera) arose a vast multitude of descendants. Modern tropical forests are home to possibly tens of millions of beetle species, and some published estimates suggest that there may be as many as thirty million to fifty million, an overwhelming number that has led entomologist Mark Moffett to describe earth as the "planet of the beetles."[13] To what do these insects owe their astronomical success? They alone have developed the ultimate body armor while maintaining the benefits of flight dispersal. A beetle's front wing is modified into a hard shell, known as the elytron, which covers the hind wing when it is at rest. When a beetle flies, its hind wing unfolds into one that is larger than the front, and flight is powered entirely by these extended back wings: an unusual arrangement called posteromotorism. The shell-like front wings are held outstretched and can only generate lift, glider-style.

During the Late Permian there were only a few groups of beetles, about six families, and they all belonged to the most primitive suborder of beetles. They were the first wood-boring insects, living in the forest undergrowth where they buzzed and flew from one fallen dead tree to another. Their hard armored bodies protected them from insect predators while they chewed into damp decaying wood to lay eggs; this environment sheltered their larvae, wood-boring grubs, from dry air and sunlight. The beetles were among the first organisms to feed on lignin and cellulose by mixing wood with fungi.

FIGURE 6.3. A beautifully preserved fossil of *Liomopterum ornatus* (family Liomopteridae) from Permian rocks of Kansas with well-developed paranotal lobes on the first thoracic segment. This neopteran (new-winged) insect family was another casualty of the Permian extinctions. Formerly placed in the Protorthoptera, these insects are now regarded as likely stem-Plecoptera. While they became extinct, their aquatic, stream-dwelling stonefly relatives (order Plecoptera) survived and flourished. (Photo by Frank Carpenter. Museum of Comparative Zoology, Harvard University. © President and Fellows of Harvard College.)

Bark lice (order Psocoptera) joined the first beetles in the deadwood. Tiny voracious insects with chewing mouthparts and gradual metamorphosis, bark lice gnaw on organic materials under the loose bark of dead trees and can congregate in massive numbers. Their modern cousins, the book lice, will feed on paper, and if undetected they can completely destroy library books. During the Late Permian, bark lice and the first beetles teamed up with wood roaches and fungi to help quickly decompose and recycle nutrients from dead forest trees and leaf litter.[14]

One Main Suspect?

Many insect species may have gone extinct during or near the end of the Permian, but virtually all the orders with complete metamorphosis survived, as well as many others with gradual metamorphosis, such as bark lice, thrips, and homopterans. Only one order with complete metamorphosis, the tiny and scarcely known Miomoptera, vanished then. Two other small groups, the little-known orders Glosselytrodea and Paratrichoptera, endured beyond the Permian–Triassic boundary and became extinct much later, during the middle Mesozoic era. Maybe they were not able to adapt to the Mesozoic's environmental changes. Maybe they were exterminated by the warm-blooded dinosaurs. Maybe they were failures, or maybe they just evolved into more modern groups. Whatever happened to the Glosselytrodea and Paratrichoptera doesn't really matter here. The key point is that they survived the Permian.

So did a lot of the other orders, which went on to diversify and are now common. Some insects, like the Homoptera, Neuroptera, Coleoptera, and Mecoptera, enjoyed substantial species richness in the Late Permian, suffered some declines, but carried on successfully into the Triassic years. Other groups, the Trichoptera, Lepidoptera, and Diptera, had low diversity but nevertheless survived the end-Permian holocaust. They are currently among the most species-rich orders. Why did they survive? The idea that low-diversity groups, like the trilobites, are particularly prone to extinction does not seem to apply here. Some of these groups could have lived on happily, provided that they had pioneered ecological niches in unaffected habitats and had an arsenal of survival skills, like complete metamorphosis. The suspect in the Permian killings must be some kind of selective agent. We are looking for a killer that could wreak havoc on coral reefs and massacre the coastal lowlands but leave the upland communities in comparative bliss.

Perhaps one factor ties together the multifarious elements of this story. We still need to consider our old friend plate tectonics, also known as continental drift. You may have already read or learned about continental drift,[15] and you may recall that before the present continents were configured, there was a time during the Early Mesozoic when the northern continents were joined in a landmass called

Laurasia, and the southern ones in a landmass called Gondwana. You may recall also that prior to that time all the land areas were united in a single vast supercontinent called Pangaea. Present-day North America was wedged directly into South America, Africa, and Europe, and present-day Africa, most centrally located, was directly connected with North America, Europe, Asia, South America, Antarctica, India, Saudi Arabia, and Australia. Although many people mistakenly assume that this massive land aggregation was the starting point for continental drift, that's not the case at all. It's just the origin of our modern arrangement of continents. Pangaea first formed during the Early Permian, when more ancient continental configurations aggregated, and after parts later broke away, it reformed again in the Middle Triassic. It took millions of years to assemble and corresponds suspiciously closely to the end-Permian extinction event.

As islands and continents collided, previously separated areas were merged. This would have reduced marine areas by eliminating shorelines and their wetland communities, and it would have brought different communities of plants and animals together. When these groups mixed, competition would insure that some species would dominate and become "weedy" and widespread, while other, less-aggressive species that formerly survived in isolation would be driven to extinction. As larger continental masses collided and fused together, these processes would have accelerated. Volcanic eruptions might have been triggered; inland, new mountains would have been uplifted and new rivers and streams would have emerged, ripe for insect colonization. As the land area became larger, the global climate changed. Inland areas became hotter and drier. Some plants and animals, like the insects with complex metamorphosis, were better able to adapt to these shifting environments.

Can Pangaea alone explain the Permian extinctions? As compelling as those arguments sound, the current answer is "no." In past decades we used to think that the extinctions took place over millions of years, but careful studies of Permian period sediments in China by Douglas Erwin and other scientists has narrowed the interval of the end-Permian extinction down to a mere hundred thousand years or less. That may seem like a long time, but it is way too fast for plate tectonics alone to have been the main culprit. However, I still think it's important to recall that they may well have been a key factor in many of the terrestrial insects' successful diversification. Pangaea's forma-tion helped bring about the Permian's arid terrestrial climate, which is usually tagged as the reason for the holometabolous insects' rapid evolution then.

The end-Permian extinctions don't look like a single event or a fast event caused, for instance, by an asteroid impact. Cosmic collision en-thusiasts have been searching the planet for geological evidence for more than thirty years, but they have found nothing definitive. They haven't identified an impact crater or impact debris from this time period. Moreover, the Permian–Triassic boundary layer lacks iridium, which is contrary to what is expected for an asteroid impact. Claims of the discovery of fullerenes ("buckyballs") in the Permian–Triassic boundary layer have filled collision enthusiasts with excitement, but these reports have been contested and are unrepeated. Impact sup-porters have even gone so far as to suggest that an asteroid collision may have triggered the Siberian volcanic events, which in turn oblit-erated the impact crater. That's a really sexy idea, but it still lacks good evidence. The philosopher Alfred North Whitehead has advised us to "seek simplicity and distrust it." His advice seems good in this case. We will keep seeking a simpler explanation for the end-Permian extinc-tions, as that is the nature of science, but for now they still appear to be complicated.

During the Permian years, insects suffered their most catastrophic setback of all time. In telling the tragic tale of the fall of the old-wings and gigantic air dragons, it's easy to forget the flip side of that story. Despite all the species and orders that died then, lots of others lived on, and they speak volumes about the insects' resilience and adapt-ability. Most of the survivors were species with novel adaptations, such as advanced neopterous wing flexing and complex metamor-phosis. The new-winged insects with wing-folding mechanisms could fly faster, hide better, and outcompete the old-winged insects. New feeding innovations like the invention of refined fluid feeding mouth-parts allowed insects to successfully colonize different plants in up-land areas. The evolution of larval stages enabled them to invade plant tissues and more successfully avoid large predators. Insects proved themselves resilient in the face of continental collisions, global cli-mate change, massive volcanic eruptions, and substantial geochemical changes in the oceans. In fact, despite the challenges they encountered while colonizing the land during the Silurian years, insects proved in the Permian that the conquest of land was a winning strategy. As cockroaches demonstrated to the trilobites, if there was ever a good time not to be in the oceans, it was during the end of the period.

Maybe it's perilous to divide the history of life into separate ages, but in this case perhaps we can indeed pick one singular moment when the Paleozoic era came to an end. I'd choose the particular day when the final trilobite died. What other creature better symbolizes the entire era than the trilobite? Their reign in the oceans lasted for more than 300 million years, but some 252 million years ago, on a cloudy morning perhaps, the last one stopped feeding in a shallow tidal pool. Her body floated to the surface, and the retreating tide washed it ashore along with other trilobite carcasses. Where were the black birds of death? Who witnessed that event? There were no shore birds on that lonely beach, but there was a scurry of small feet, as first one cockroach, then another, found the castaway body and consumed it. Maybe a lone beetle, preening its antennae on a log nearby, briefly flew down to inspect the scene and partake in the feast. Then it turned, unfolded its wings, and buzzed clumsily into the forest. The Paleozoic years had ended, to quote the poet T. S. Eliot, "not with a bang but a whimper," and the insects declared themselves the victors. But soon there would be a new rustling in the Mesozoic forests; with slashing claws and gnashing teeth, the first dinosaurs would appear. How would the insects get along with their new roommates?

PLATE 1. The disruptive coloration of the malachite butterfly, *Siproeta stelenes*, provides excellent camouflage in the dappled understory of the tropical dry forest at Chamela Biological Station in Jalisco, Mexico. Examples of the largest butterfly family, Nymphalidae, they fly rapidly when disturbed.

PLATE 2. *Facing page, top.* Can you see it? Motionless on a leaf, this Ecuadorian katydid nymph (order Orthoptera, family Tettigoniidae) displays remarkably cryptic green coloration. (Photo by Angela Ochsner.)

PLATE 3. *Facing page, bottom.* Stalking the mossy Ecuadorian cloud forest at night, this well-camouflaged praying mantis nymph (order Mantodea) uses crypsis to its advantage while hunting insect prey. (Photo by Andy Kulikowski.)

PLATE 4. *Above.* Hiding in plain sight, this very cryptic katydid demonstrates another method of camouflage: by resembling tree bark it is well adapted to surviving the prolonged dry season at the Chamela Biological Station in Jalisco, Mexico.

PLATE 5. *Above*. Seemingly confident in its ability to rapidly jump away from danger, this Ecuadorian short-horned grasshopper (order Orthoptera, family Acrididae) is sitting pretty at the Yanayacu Biological Station in Ecuador. (Photo by Angela Ochsner.)

PLATE 6. *Facing page, top*. Voracious predators of other insects, the assassin bugs (order Hemiptera, family Reduviiidae) impale their prey with their piercing mouthparts and suck out the fluids. (Photo by Angela Ochsner.)

PLATE 7. *Facing page, bottom*. Literally stuck in one place, the sap-feeding spittle-bug nymphs (order Homoptera, family Cercopidae) cover themselves with bubble-laced honeydew. This Ecuadorian species may acquire defensive chemicals while feeding, since it also displays bright aposematic warning coloration. (Photo by Angela Ochsner.)

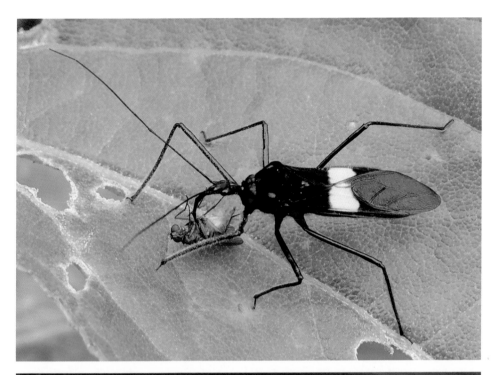

PLATE 8. *Facing page, top.* These *Altinote neleus* butterflies (family Nymphalidae) commonly gather roadside at mud puddles on sunny afternoons at the Yanayacu Biological Station. Perhaps overly confident in their toxic defenses and aposematic coloration, they can be picked up by hand and will exude orange toxic fluids from their bright abdomens. Although they are well evolved against predation, they have not adapted to the modern world and are often run over by passing vehicles. (Photo by Angela Ochsner.)

PLATE 9. *Facing page, bottom.* Seldom seen by daylight, the enormous tusk-jawwed males of *Corydalus hageni* (order Megaloptera, family Corydalidae) sometimes fly to lights at the Yanayacu Biological Station.

PLATE 10. *Above.* A pair of brightly colored tortoise beetles (order Coleoptera, family Chrysomelidae) display just the right combination of body armor, defensive chemicals, and aposematic coloration to leisurely feed on leaves in the Ecuadorian cloud forest. The chrysomelid leaf beetles are a highly successful group of plant feeders in the New World tropics. (Photo by Angela Ochsner.)

PLATE 11. *Top.* Careful handling this one! The caterpillar of a hemileucine saturniid moth (family Saturniidae) displays its defense mechanism: brittle, hollow spines loaded with irritating toxic chemicals. This approach works well to protect these caterpillars from predation by birds and other vertebrates but does nothing to defend them from parasitism by small wasps and flies. (Photo by Jennifer Donovan-Stump.)

PLATE 12. *Bottom.* Large adults of *Automeris abdominals* are frequent visitors to the lights at the Yanayacu Biological Station. By day they rest with their brown forewings covering their hind wings, rendering them very cryptic on leaf litter. When disturbed they reveal the bright hind wings with large eye spots, which may induce a startle response in predatory birds. (Photo by Angela Ochsner.)

7 Triassic Spring

April is the cruelest month, breeding
Lilacs out of the dead land, mixing
Memory and desire, stirring
Dull roots with spring rain.

T. S. ELIOT, *The Waste Land*

It's difficult to judge the seasons in Wyoming. In June, sporadic rain transforms the prairie grassland into an endless vista of emerald green, which waves in the wind like a vast ocean. At higher elevations the mountain meadows are awash with red, yellow, and blue wildflowers, which overnight might be buried under a snowfall. By July, the summer drought quickly dries out the grasslands and flowery fields, and the prairies and meadows turn golden brown for another ten months. Even so, in mid-July, it's not unusual for a hailstorm to cover the ground with ice pellets.

The trees unfortunately don't help much. Deciduous ones like aspens and poplars don't leaf out until late May or early June. Autumn comes early. The first killing frost usually occurs in mid-September, and the fierce Wyoming winds quickly strip the trees of dead leaves by early October. The broadleaf trees tend to be without leaves for a full eight months. If you judge the winter by when the trees are bare, then it lasts a long time.

It's not so easy to judge by snow, either, which at high altitudes can come during any month. Our heaviest storms tend to arrive very early in the fall or late in the spring, burying Wyoming in white crystal flakes in October or May. If that's not confusing enough, some of the most pleasant weather is liable to show up in midwinter. On most days in January or February, the skies will be totally clear and luminous azure blue. The winds tend to blow away the snow, or it desiccates in the dry air, so long stretches of winter may be totally without any snow cover.

Anyone who has lived in Wyoming can probably understand T. S. Eliot when he wrote that "April is the cruelest month," as the onset of spring is particularly hard to judge. On sunny days, tulips and daffodils rapidly shoot up, usually to get frozen in ice or buried under snow. There are two other indicators of spring aside from the first flowers, and they're especially significant because they remind us of the Triassic years, 252 to 201 million years ago. The first is the traditional one: the appearance of robins and other migratory birds. In my home state of Michigan, the robin is indeed a pretty good sign of mild spring weather. Wyoming is crueler: more often than not, this bird appears all puffed up, and desperately tries to keep warm by hiding among the snow-laden spruce boughs during a late spring blizzard. But why would robins remind us of the Triassic?

When I was a child, birds were birds and dinosaurs were thought to be great reptiles. How times have changed. We now understand the message of the wishbone: birds are direct descendants of the dinosaurs, the most famous of all groups which originated during the Mesozoic era.[1] The dinosaurs first evolved during the Triassic years, and the feathered dinosaurs and first birds arose during the Jurassic; while the big dinosaurs like *Tyrannosaurus* and *Triceratops* went extinct at the end of the Cretaceous, the little feathered ones flew on into history. More species are alive now than during Mesozoic times. They just have feathers, wings, beaks, and no teeth. We should remember the dinosaurs every spring when we see the first robin.

But in Wyoming the migratory birds can be unpredictable. In some years they are late arriving, and in others they seem to bypass our area entirely. So I've found another indicator of spring, a resident animal that's a sure sign every year: xyelid sawflies, which live at high elevations near conifer trees and willow bogs. In the early spring, color returns to the willow twigs, providing bright yellow, orange, and red twiggy contrasts against the brilliant white snowbanks melting along the hillsides. Soon the willow buds swell and burst into a profusion of furry pussy willows. Once they start producing pollen, some very tiny insects arrive to gather a high-protein meal. Among them are the xyelid sawflies, which after emerging from their overwintering cells in the soil are usually the first adult insects to become active in the Wyoming mountains.

Xyelid sawflies are living fossils, relicts of the Triassic years. They are the most primitive living group of the insect order Hymenoptera, the lineage that now includes some of the most successful of all insects: bees, ants, social wasps, and parasitic wasps. And what curious creatures they are. The base of their antenna and maxillary palpus (a segmented feeding appendage located behind the mandible) are leglike, giving them the appearance of having extra legs near their front end. They use their leggy mouthparts to gather a meal, and then they fly up to the tops of nearby conifers, just as they did millions of years ago. The name "sawfly" refers to the females' serrated egg-laying organ, what entomologists call an ovipositor. With it the females abrade slits into developing pollen-bearing pine cones and insert eggs into the nutritious plant tissue. After the eggs hatch, the young xyelid larvae feed on the tissue, drop to the soil, and finally dig cells in which they pupate until the following spring. This habit served the ancient sawflies well, as it allowed their larvae to live in the treetops, where they were able to evade even the tallest of the dinosaur macroherbivores: the brontosaurs.[2]

Breaking the Silence of Permian Extinction: Triassic Not-So-Silent Spring

What period of life could be more springlike than the Triassic? After the catastrophic end-Permian mass extinction came a time of rebirth. The landscape was drier than during the Paleozoic years, but the Triassic forests became dominated by new kinds of plants: mostly conifers, cycads, gingkoes, and ferns. Triassic conifers are now famous because some of them fell into rivers and were washed into coastal lowlands, where they were buried in volcanic debris. Over time these trees fossilized into colorful quartz minerals and became the petrified trees located across the southwestern United States, especially in Arizona's Petrified Forest National Park.

There were no flowering plants, fruit, berries, or grasses in the Triassic. Still, with the prevalence of conifers and the loss of the ancient insect orders, Triassic forests would look familiar to us. Myriad aquatic insects, including mayflies, damselflies, stoneflies, and cad-

FIGURE 7.1. A xyelid sawfly, *Pleroneura californica*, a sure sign of spring in Wyoming, is a relict of an ancient insect family that arose during the Triassic period.

disflies, fluttered along the streams. The forests buzzed and hummed as legions of new species appeared, especially among the roaches, crickets, planthoppers, true bugs, lacewings, beetles, scorpionflies, and true flies. The Triassic also saw the origin of several new insect groups, some unfamiliar and some now very familiar, such as the walking stick insects, webspinners, earwigs, dobsonflies, snakeflies, and wasps. Vertebrates were busy snapping at all these creatures: the first turtles, salamanders, frogs and of course, most notably, the very first dinosaurs had arrived. By the Late Triassic, several dinosaurs roamed the forests, including some very small bird-sized species like *Saltoposuchus* and *Procompsognathus*, and a few larger ones like *Plateosaurus*.

You've no doubt heard lots about the dinosaurs' impressive dynasty. They certainly became the dominant macroherbivores and macropredators of the Mesozoic era, fiercely overshadowing the mammals for a hundred million years or more.[3] But I'll put a twist on the story: dinosaurs were impressive characters in a big Mesozoic world already largely filled with insects. We don't usually hear much about the both

FIGURE 7.2. A beautifully preserved fossil dragonfly, *Protolindenia wittei*, from Jurassic rocks of Bavaria. The aquatic insect order Odonata (dragonflies and damselflies) is one of only two paleopteran (old-winged) insect orders which survived the Permian extinctions and flourished in the Mesozoic era. (Photo by Frank Carpenter. Museum of Comparative Zoology, Harvard University. © President and Fellows of Harvard College.)

of them together, but dinosaurs must have influenced the insects, and I'm sure the insects affected the dinosaurs.

People are still debating what the earliest dinosaur was like, but a popular notion is that the South American "rabbit crocodile," *Lagosuchus*, might be the very first one—or certainly an early model suitable for illustrating the dinosaurs' origins. *Lagosuchus* wasn't, of course, a rabbit or a crocodile, but an honest-to-goodness dinosaur with the group's defining feature: a long scimitar-like thumb-claw that could swing in two directions. Like crocodiles, they were long, slender, graceful, and they had an extended narrow snout and lots of sharp teeth; like rabbits, they had long, powerful hind legs, which allowed them to run and jump. By some accounts they could climb trees and probably hop from tree to tree.

According to the orthodox view, the very first dinosaurs were carnivorous predators. Because of their sharp teeth and flexible thumb-claw, which was good for slashing and ripping, we can safely assume that *Lagosuchus*, for instance, was a meat eater. Then along came *Plateosaurus*, the first of the plant-feeding dinosaurs. The plateosaurs could reach up into the lower branches, grab them with their thumb-claws, and pull them low enough that, reaching with their long necks, they could eat the tasty leaves. Yes, bloody red meat and green leafy salad, that's what the dinosaurs craved and ate. That's what all the good books tell us. What more could they want?

It's probable that dinosaurs craved dietary diversity. Maybe they had enough sense to improve their nutrition by consuming a wide variety of small tasty items, like soft-bodied insects that would be fully digested and not show up well in dinosaur coprolites. Virtually all living vertebrate animal groups feed extensively on insects. Fish do—at least freshwater fish eat lots of bugs—and so do amphibians, reptiles, birds, and most mammals. Early mammals were insecti-vorous, and so were early primates, especially lemurs. Many human cultures even include insects as part of a broader diet because they are an excellent source of protein, fat, calories, several trace minerals, and B vitamins.[4] How can we suppose that only dinosaurs, among all the major vertebrate bloodlines, ignored them? Small predatory dino-saurs and the young certainly wouldn't have turned up their noses at an insect meal. But even the plant feeders must have eaten lots of bugs, either accidentally or intentionally. Some dinosaurs might have pref-erentially selected plant parts with the most edible insects because, like bacon bits in a salad, they would have provided more nutrition than just conifer needles and cycad fronds alone.

Let's think for a moment about the kinds of insects modern verte-brates like to eat. Mostly they fall into one of two categories: relatively large insects, like fat wood-boring beetle larvae or shrimp-sized cater-pillars, or much smaller ones that can be located easily in large num-bers. These might be swarming species that are super-numerous at certain times or gregarious species that live in large groups. During the Triassic, plenty of juicy wood-boring beetle grubs were chewing in fallen trees, large cicadas were tippling on forest vegetation, and other insects were feeding in leaf litter and soil or tunneling in plants. Dino-saurs had eyes and ears, of course, so they could see the insects moving on surfaces and hear them chewing through decaying logs. And their nimble thumb-claw was a perfect tool for digging grubs from rotten wood. Surely the early small dinosaurs hunted for beetles there, just like woodpeckers and long-clawed insectivorous primates. Also, even though the social insects didn't exist in the Triassic, plenty of swarming species would have provided the first dinosaurs with seasonal feasts. While at rest, they need only have used their tongues to lap these insects off foliage. Then again, the earliest dinosaurs certainly had the ability to jump at flying insects and snatch them from the air. The homopterans, which by the end of the period had more species richness than any other insect order and were super-abundant, also provided a likely meal. Because of their piercing mouthparts, planthoppers are literally stuck in place while they are drilling for food. For the dinosaurs, they would have been easiest to find.

Insect diversity must have shaped dinosaur diversity, not only because various small and herbivorous dinosaurs likely depended on bugs for protein, but also because the carnivorous dinosaurs ate an assortment of tiny amphibians, reptiles, and mammals, all of which were largely insectivorous. In return, dinosaurs likely shaped the evolutionary patterns of some insects. Although the Triassic ended long before the advent of alkaloid-bearing flowering plants, ancient species had various secondary chemical defenses, which some Triassic plant-feeding insects might have adopted to fend off dinosaurs. Moreover, these insects might have evolved aposematic warning coloration—bright yellow, orange, and red—since modern insects with chemical defenses tend to evolve this kind of protection when visually searching predators feed on them. Others might have evolved cryptic colors that resembled plants, wood, or soil. And some might have followed a different route altogether, developing behavioral escape mechanisms; it's certainly possible that the mayflies' and cicadas' mass synchronized emergences adapted and were fine-tuned in response to intense dinosaur predation. In time, Triassic insects exploited the dinosaurs: there emerged dung-feeding beetles that harvested dinosaur droppings as well as blood-feeding and, possibly, parasitic insects, which might have fed directly on them.

The Dinosaurs' Buggy World

Several important insect orders survived the Permian, eleven of which thrived during the Triassic years and still survive today: Ephemeroptera (mayflies), Odonata (damselflies), Blattaria (roaches), Orthoptera (crickets), Plecoptera (stoneflies), Homoptera (planthoppers), Neuroptera (lacewings), Coleoptera (beetles), Mecoptera (scorpionflies), Trichoptera (caddisflies), and Diptera (true flies). More than 90 percent of the Late Triassic insect species belonged to these eleven orders. Moreover, at least eight new orders first appeared during the Triassic. Seven of them seem more or less familiar because they also still exist: Phasmatodea (stick insects), Embiodea (webspinners), Dermaptera (earwigs), Hemiptera (true bugs), Megaloptera (dobsonflies), Raphidioptera (snakeflies), and Hymenoptera (wasps). But one order of new insects might legitimately be considered the most distinctly Triassic because it lived only during the period. These are the giant titan insects of the order Titanoptera.

The titans were large orthopteroids with sharp chewing mouthparts, and they looked a bit like oversize katydids. They had spiny front legs suitable for grasping victims, and they are thought to have preyed on other insects. The largest of the titans had wingspan of thirty-six centimeters (more than one foot wide) and were large enough to catch and eat small vertebrates like salamanders and frogs. The males had large, reticulated sound-producing areas on their front wings, and during the mating season, they must have produced loud songs as they called for mates and defended territory in the forest vegetation. No one knows why the titans disappeared after the Triassic times, but their extinction corresponds closely with the diversification of small dinosaurs and the appearance of the first birds. So maybe the titan insects were just too big to be that noisy and get away with it. Perhaps they were the first casualties of the emerging brigade of insectivorous feathered dinosaurs.

The phantom insects of the order Phasmatodea (also known as Phasmida)—or as we commonly call them, the stick insects and the leaf insects—evolved a highly successful strategy for surviving in a forest filled with hungry dinosaurs: crypsis, the development of forms and colors that provide camouflage. Some lost their wings and developed long, thin bodies resembling sticks, while others evolved green

FIGURE 7.3. A very cryptic stick insect from Ecuador, an example of the insect order Phasmatodea, which began its successful rise in the Triassic period. (Photo by Angela Ochsner.)

wings that resemble leaves. All of these insects chew on plant tissues, and they further avoid predators by mostly feeding at night. Phasmatodeans didn't evolve crypsis because they wanted to look like plant parts. Rather, crypsis is the result of natural selection enforced by visually searching predators. Those phasmatodeans with the most convincing forms and colors survive to reproduce, while those with less-effective colors tend to get eaten more often. Notably, stick insects first evolved in dinosaur-ridden tropical forests even before the birds appeared. Maybe the wild diversity of phasmatodean body forms in the modern world is, in part, a legacy of dinosaur feeding habits.

If the dinosaurs tilted a curious eye at stick insects from time to time, they surely must have seen another mystifying creature. During the Late Triassic, layers of white silk started to cover tree trunks, rocks, and leaf litter. These weavings were the handiwork of the elusive order Embiodea, also known as webspinners. The lively webspinners evolved their own strategy for totally avoiding dinosaurs. They developed very small body sizes, invented a whole new way of making silk, enveloped their world with the material, and stealthily went into hiding, living peaceably for 220 million years by chewing on lichens and decaying vegetation. Instead of spinning silk from their mouth or anus, the webspinners spun it from a novel silk gland in their front legs while they walked. Because of this unusual habit, I like to point out that they are the insects most like Spiderman: they are the only ones to use their "wrists" like Peter Parker. Heck, even spiders don't do that. Spiderman really should have been called the "Webspinner" or "Embiidman."

Scurrying in the leaf litter, with long pincerlike structures on their tail ends, the earwigs joined the webspinners. These are the "skinwinged" insects of the order Dermaptera—and the origin of their name is obscure. Legend has it that during the Middle Ages, earwigs used to crawl into people's ears or wigs while they were sleeping. That may well have been the case, as earwigs are nocturnal scavengers that feed mostly on decaying vegetation then look for dark hiding places in the morning. These pinching little insects are so annoying that I'm going to speculate that even the dinosaurs didn't like them much.

I imagine that a morning in the life of the little Triassic dinosaurs looked something like this. After a nice relaxing sleep, they rose early as the sun's rays slanted through a dense forest cover of cycads and ferns and got rid of their earwigs. After quickly looking around, they probably started the day by trotting down to a nearby stream for a refreshing drink of cool water. Then maybe after picking a few newly emerged aquatic insects off the streamside vegetation, they began to forage and tear apart some rotten, fallen trees by the water's edge in search of yummy beetle grubs. There were plenty of grubs to satisfy the heartiest appetite, but sometime during the Late Triassic the dinosaurs started encountering the pupae of yet another new insect in the pulpy streambed logs, large and tasty ones that would not go down so easy because they could bite back. These pupae of the dobsonfly are notable in being the only pupae that have functional mandibles, with which they are able to defend themselves. When the adult dobsonfly

FIGURE 7.4. This webspinner from Thailand, *Eosembia auripecta* (family Oligo-tomidae), is an example of a successful but cryptic insect order that evolved during the Triassic, alongside little dinosaurs. This tiny female is only about 1.8 centimeters long. (Photo by Janice Edgerly-Rooks.)

emerges from its pupation chamber, it has very large wings; hence the scientific name Megaloptera, which means large-winged insect.

But the dobsonfly's wings are not its most impressive feature. Surely the males' jaws are, even to the eye of a young dinosaur. These jaws are long and sicklelike, and they look like the tusks of a wooly mammoth. Oddly, male dobsonflies don't use them much for defense, but mostly for grasping a female during courtship and mating and for asserting their territory when confronted by rival males. It's the female who can deliver a surprisingly powerful and painful bite with her much shorter jaws. After mating, the female dobsonfly lays her eggs on the underside of a fallen tree or other object overhanging a stream, so her young, when they emerge, can fall directly into the water. The larvae aggressively prey on other aquatic insects, as well as small fish and small amphibians. They grow into large larvae, called hellgrammites, which also deliver a painful bite. Other than some paper wasp stings and ant bites, a hellgrammite bite is one of the most intense experiences insects have to offer. I know because I've been bitten by one, and I suspect that as the hellgrammites crawled out of water in search of pupation chambers, the foraging dinosaurs found them daunting and learned to avoid them.

Real Bugs Don't Eat Quiche

When dinosaurs visited the streambeds for a morning drink, they might also have paused to observe some tiny insects moving along the moist sand near the water's edge. They probably wouldn't have bothered to eat these insects, which were mostly too small, but I suppose the dinosaurs might have regretted not finishing them off in the Triassic. The first of the insect order Hemiptera, the group that entomologists give the fine distinction of being called "true bugs," this bloodline later spawned bed bugs, assassin bugs, and giant water bugs—all of which might have been capable of biting dinosaurs before the Mesozoic was over. The true bugs are more familiar to us as the pleasantly scented stink bugs, plant bugs, and seed bugs, among others, but these are all much later species.

The Hemiptera first arose as a lineage of semiaquatic predators and beachcombers. Like their closest cousins, the plant-feeding homopteran planthoppers, true bugs have liquid-feeding, piercing mouthparts. Unlike their cousins' beaks, however, the beak of a true bug is on the front of its face, and it's longer and more flexible; with it true bugs started sucking the blood out of other hapless insects. True bugs had a very auspicious start, for few insect groups have diversified so greatly in terms of the new habitats they later colonized. From streambeds they moved to the water's surface and then into the water, where most remained fully aquatic predators; a few, however, became scavengers of green algae and plankton.

Later in the Mesozoic era, one group, the water skaters, learned to walk on water more efficiently than any other animal. They are the only insects that moved back downstream and recolonized the open oceans; other true bugs moved from the waterside back into the forests. While some became specialized plant feeders alongside their relatives, the planthoppers, others, like assassin bugs, became violent predators of other forest insects. Some of these assassin bugs were blood feeders who feasted on small mammals, and they would have bitten a dinosaur, given the chance.

Back in the forest, up in the conifers, emerged the snakeflies (the insect order Raphidioptera). I don't suppose the dinosaurs down below would have noticed these hidden dwellers in the treetops: they probably minded their business and stayed out of the dinosaurs' range. Most people also never see snakeflies, as they are rare worldwide, but oddly enough they're another Triassic insect that lives on and is common here in Wyoming. We call them snakeflies because the adults have a long neck and prey on other insects, and when they attack they strike their head forward like a snake. Modern snakeflies feed almost entirely on homopteran insects like aphids; their Triassic ancestors probably did much of the same thing, moving about in the treetops and mowing down herds of defenseless planthoppers. The female has a long, curved, taillike ovipositor with which she lays her eggs on plant surfaces, where the larvae hatch and feed like the adults. Between its long neck and ovipositor, the snakefly's profile looks more like a brontosaur than a snake. Given the context in which they evolved, maybe we should call them brontosaurus bugs, because over the duration of the Jurassic years, while Wyoming was still a coastal lowland tropical forest, the snakeflies fed in treetops alongside nibbling brontosaurs.

The xyelid sawflies, first of the insect order Hymenoptera, also evolved up in the Triassic treetops. I mentioned them early in this chapter not just because they're a sure sign of spring but because of all the Triassic insects, they alone would blossom into a massive dynasty of millions of species. "Hymenoptera" means "membrane-winged," but there is a legend about this name. In Greek mythology, Hymenos was the god of marriage; some say the name Hymenoptera derives from the joining, or marriage, of hymenopteran wings during flight by a row of tiny hooks called hamuli. These wings were helpful because they allowed the sawflies to move from their pupation chambers in the soil up into the treetops. But their sawlike ovipositors proved all the more useful, as they allowed the sawflies to place their young deep inside plant tissues, where they were safe from snakeflies that fed only on the surface.

From those ancestral sawflies flitting about in the treetops ultimately arose diverse bloodlines of social insects: bees, ants, and social wasps, as well as nest-provisioning stinging wasps and an extraordinary array of parasitic wasps.[5] Over the long run, the ovipositor was more important than wings in the rise of the wasp dynasty. Many modern wasps have lost their wings, but the ovipositor has become modified for multiple uses: for placing eggs into plant tissues, onto other insects, or deep inside other insects' bodies. Perhaps most significantly, the ovipositor has become modified into a stinger used for both paralyzing prey and defending against predators. It sure didn't look like much in the Triassic years, but by the Cretaceous that little sawfly dagger would transform into a toxin-filled hypodermic needle fully capable of inflicting pain on even the largest and fiercest dinosaur.

Perhaps the most unique feature of sawflies and wasps is something that's not externally visible: the sex of each individual is determined by whether or not an egg is fertilized. An unfertilized egg develops into a male. If an egg is fertilized, it always develops into a female. When wasps mate, the male transfers sperm cells into the female's sperm storage organ, the spermatheca, where they remain viable for a long time. Fertilization does not take place during mating, but rather is done by the female at the time an egg is laid, allowing her to choose the sex of each of her progeny. Some ecologists suggest that this allows female wasps to assess the quality of available food and selectively place female eggs on the best items. Also, when males are scarce, it allows females to lay eggs that develop only into males— a very useful trait allowing these insects to exist in highly dispersed populations. This curious method of sex determination may also have profoundly impacted the evolution of social behavior in wasps, ants, and bees. Oddly enough, it created a situation where sisters are more closely related to each other than they are to their own daughters. A worker bee or ant, for example, actually passes along more of her own genes to the next generation by helping the queen (her mother) raise more of the queen's offspring, than if that same worker had her own daughters.

The two largest wasp family groups, Ichneumonidae and Braconidae, each contain more species than all of the vertebrates combined, and in tropical areas the social insects surpass the vertebrates in total biomass. Wasps are keystone predators and parasites of most plant-feeding insect populations; consequently, they greatly influence the diversity and abundance of forest plant species. Ecologically, wasps are also important as herbivores, seed dispersers, pollinators, and nutrient-recyclers. They probably haven't had the good press they deserve, but the wasps are the grandest success story of the Triassic years, even more so than the dinosaurs.

The Late Triassic, about 201 million years ago, was a tough time for mammals nuzzling around in the leaf litter, searching for a meal. But it was a prime time for dinosaurs, and a glorious one for bugs. It would probably have been difficult to spot a dinosaur in the Triassic forests, just as it's difficult to spot large vertebrates in modern tropical forests, but those areas were saturated with insects. While the shorelines teemed with aquatic species, the leaf litter crawled with an assortment ranging from bristletails to beetles. The plants, still mostly conifers, cycads, ferns, and gingkoes, were chewed or sucked upon by myriads of orthopteroid insects, thrips, planthoppers, and a few primitive moths and sawflies. All of these were attacked by a host of predatory insects, including the titans, lacewings, scorpionflies, and snakeflies.

As the sun set each evening, the titans' baritone rasping was the loudest noise in the forest. But down below the dry leaf litter rustled as small dinosaurs prepared to settle in for a long night's sleep. Some of those dinosaurs started ruffling their feathers, which they had evolved to keep warm on long cool nights. For the first time in almost 150 million years the vertebrates were poised to start chasing the insects into the air. The Jurassic times were approaching, and the songs of the titans were about to be replaced by birdsong.

8 Picnicking in Jurassic Park

Reverie is normal in Wyoming at sunrise.

ROBERT T. BAKKER, *The Dinosaur Heresies*

Early on this sunny winter morning, as I write these words, I'm pondering two things: a riddle and the view from my window. At first glance, the two seem totally unrelated, but nothing could be further from the truth. In fact, the view first caused me to ponder the riddle, and, as you will see in a bit, both are woven into the story of the Jurassic period, 201 to 145 million years ago, when insects and dinosaurs diversified like never before, the birds first appeared, and vertebrates finally chased insects into the air.

First, the riddle: what is the largest organism an entomologist has described and named? An entomologist, you will recall, is a biologist (like me) who happens to study insects. More precisely, I examine insect evolution and classification and am involved in the discovery, description, and naming of new species: a branch of the science known as systematic entomology. The largest species I've named, a megalyrid wasp from Papua New Guinea, is slightly less than sixty millimeters long, if you count its ovipositor along with its body. Most of my new species measure only a few millimeters in length. Keep in mind while you ponder my riddle the fact that we bug hunters typically discover very small organisms. Now I'll go on to describe the view from my window.

The University of Wyoming Insect Museum is situated on the fourth floor and has broad windows facing directly north. On any given day they provide an impressive panoramic view above and beyond Laramie's northern limits. What lies beyond the city is of particular interest here: at first glance, apparently nothing. The Laramie Valley is a treeless short-grass steppe. The terrain is open, windswept, barren, and seemingly hostile; for miles nothing obstructs the horizon, which, although distant, is sharp and clear. The air is cold and crisp, the brownish tan grasslands illuminated by the sun and starkly contrasting with the crystal blue Wyoming skies. It's no problem to see all the way to the northwestern horizon and easy to imagine you can see beyond it, clear to Como Bluff, Sheep Creek, and the town of Medicine Bow. That's where our story continues, because out near Sheep Creek more than a hundred years ago, an unusually large creature was discovered—the very organism that's the object of our current riddle. How did this exceptional animal come to be in the windswept basin north of Laramie?

How you react to this puzzle may depend first on how you define the word "large." If you mean heavy or massive, then adult Goliath beetles of central Africa and larval South American Hercules beetles would be among the largest.[1] If instead you measure body length, then you might select the Australian titan stick insect (*Acrophylla titan*), which grows up to ten inches long and has wings that spread equally wide. Although longer than any other living insect, including the Goliath and Hercules beetles, an individual titan stick insect does not weigh nearly as much. To complicate matters further, one could measure size from the other dimension: width. The huge atlas moth from India (*Attacus atlas*) can have a wingspan, from tip to tip, of nearly twelve inches. The great owlet moth from Central and South America (*Thysania zenobia*) sometimes has a wingspan that exceeds that of the atlas moth, but the total surface area of its wings is less. All these insects are notably large in their own way, but all are tropical animals found nowhere near Wyoming.

Most of you have probably remembered our previous discussion of gigantic Carboniferous and Permian insects and already realized that the answer to my riddle must lie in considering not only living organisms but fossilized ones as well. How about the most majestic of all Permian insects, the nearly three-foot meganeurid air dragons? One of the larger species was indeed discovered much closer to home, in Permian rocks in Kansas and Oklahoma. But alas, the organism I have in mind is even larger—orders of magnitude larger than the meganeurid air dragons. I suppose now would be a fair time to reveal a crucial piece of information. I never did actually say that the organism was an insect, only that an entomologist named it. No rule of the Inter-

FIGURE 8.1. The South American Hercules beetle (*Dynastes hercules*) is the largest of the horned rhinoceros beetles. A large male, such as this one from Ecuador, can measure nearly seven inches long, and their larvae are among the heaviest insects known. (Photo by Angela Ochsner.)

national Code of Zoological Nomenclature says that you have to be an entomologist to name an insect, or that an entomologist can't describe organisms from any other animal group. These days an entomologist seldom does otherwise than discover insects, but in the past biologists worked more widely among diverse groups of organisms. Once upon a time, in Pittsburgh (of all places), a famous entomologist, William Jacob Holland, christened a dinosaur from Wyoming *Diplodocus hayi*. This dinosaur is not only the largest organism an entomologist has ever described, it's one of the largest land creatures that ever lived.[2]

Long-Necked Nibblers and Long Neck Biters

For most people, *Diplodocus* and its better known cousins, the bronto-saurs, are the quintessential Jurassic dinosaurs: small headed, long necked, humungous bodied, with extended whiplike tails. When I was a young child the popular dogma held that both lived in swamps, but we now know them for what they were: the heaviest land crea-tures that ever lived. If anything, they shunned the wetlands. They trotted through the tropical Jurassic forests, raising up on their hind legs and craning their serpentine necks well up into the tall trees to forage for their daily meals. A fully grown adult brontosaur (now cor-rectly called the *Apatosaurus*) could measure up to seventy-five feet long from mouth to tail tip, and it might have weighed a whopping twenty-five tons. To keep up their healthy weight, the brontosaurs had to eat an enormous amount of vegetation—by some estimates, up to a ton every day. These estimates never mention trace items, such as insects. Like any large browsing animal, brontosaurs would have selected the most-tender vegetation—the newest leaves with the most nutrition and the lowest amounts of toxic chemicals—and this vegetation, of course, would attract plant-feeding insects as well. The brontosaurs probably consumed around a couple pounds of insects along with their daily salad, assuming they just randomly nibbled on the creatures. For all we know, they might have found insect-infested foliage *tasty* and ate all they could find. Still, a ton of salad with a few insect vitamin pills is a huge meal to digest. Along with consuming forest vegetation the brontosaurs paused from time to time to swallow a few big, smooth rocks. These gizzard stones rolled around in their massive stomachs along with powerful digestive enzymes and helped them more rapidly digest their excessive meals. Just a few rocks a day and the benefits were enormous: brontosaurs needed to spend less time chewing and so could spend more time just stripping the trees of tender leaves and swallowing. These polished stones still litter the windswept plains near Como Bluff and Sheep Creek, a mute testimony to the dietary prowess of the most gargantuan feeding machines that ever wandered the earth's forests.

Allow me to rephrase that last point. The brontosaurs and diplodicines might well have been the heaviest animals to ever roam the woodlands, but the Jurassic years spawned some even more impressive feeders. Perhaps now would be a good time to debunk a few of our myths about the Jurassic. You've probably read *Jurassic Park*, seen the movie, or viewed one of its sequels. While the movies did much to improve public understanding of dinosaurs as warm-blooded, fast, and possibly intelligent, they unfortunately contributed to several major misperceptions about what life was like during the Jurassic, the most notable of which relate to the large predatory dinosaurs. Despite what the movies portray, there were no tyrannosaurs or velociraptors in the Jurassic. Sure, those animals did exist, but they lived in the Late Cretaceous—tens of millions of years later. Although they may have existed around the same time, *Tyrannosaurus* and *Velociraptor* probably never ran into each other, because they lived in different parts of the world: *Tyrannosaurus* in western North America and *Velociraptor* in Mongolia.

And while the velociraptors might have had big brains and hunted in packs, they were not nearly as sensational in reality: fossils indicate that they were only about 25 percent as large as the creatures depicted in the movie. A full-grown tyrannosaur could have stepped on a velociraptor with impunity, and if they ever did run into each other, the velociraptors probably quickly fled in the other direction.[3]

This Jurassic confusion is totally unnecessary. One Jurassic predator was, in fact, as ferocious and impressive as any tyrannosaur or velociraptor: the *Allosaurus*. Some paleontologists speculated that allosaurs might have been scavengers, but let's give them some credit where credit is due. Adults were between twenty-seven and forty feet long and might have weighed one to two tons. They had razor-sharp slashing teeth and claws and large jaws that could disarticulate like a snake's to take huge bites of meat, up to a hundred pounds at a time. You don't need teeth and claws like that to nibble on dead carcasses. Although the allosaurs might not have been as big as the tyrannosaurs, they hunted meals that were larger than anything any tyrannosaur or velociraptor ever encountered. They were the most abundant large predatory dinosaur in the Late Jurassic of North America, and the wealth of allosaur bones in the same fossil beds suggests that, like *Velociraptor*, they were social and hunted in packs.

Imagine the real Jurassic park: near a river basin thick with ferns and cycads, a herd of brontosaurs serenely browses on the high tree branch foliage of the bordering coniferous forest. Suddenly, out of the thick, shadowy undergrowth rushes a pack of snarling allosaurs, and, momentarily, confusion reigns. The dominant bull brontosaur is quick to defend his herd, adroitly whipping his huge tail, hurling a juvenile allosaur thirty feet into a fern thicket. Bellowing an alarm call, the huge brontosaur reels to the right, rears back, then stomps forward into the pathway of a charging allosaur, which screeches in agony: the brontosaur has crushed and broken the largest toe of the allosaur's right foot. The bull kicks out again but misses, and the wounded allosaur, favoring his left leg, hobbles a hasty retreat. But in the chaos of the moment the damage is done. A young brontosaur, separated from the herd, bellows a high-pitched squeal as a seasoned veteran of the hunt, the matriarchal female *Allosaurus*, lunges from behind and rips a fifty-pound chunk of flesh from its right flank. The allosaur pack retreats to the forest to lick their wounds, while the bull brontosaur leads his surviving herd to the comparative safety of a ferny vale a half-mile upriver. The mortally wounded young brontosaur, staggering from blood loss, collapses to the ground. The allosaurs have suffered three broken ribs and a broken toe but they have won the day, gaining enough meat to feed the pack for a month while they recover from their war wounds.[4]

Tiny but Mighty Jurassic Park Killers

While one can't very well look at the Jurassic years without admiring the dinosaurs, I do have an ulterior motive in bringing them up: to compare those massive meals with some tiny ones in dead fallen trees—beetle larvae in rotting wood, to be precise—eaten by wood wasps. Remarkably, 180 million years ago, the dinners of these small burrowing insects became even more significant than the gigantic meals of the allosaurs, which enjoyed the grandest predator-prey relationship in the history of the planet. To understand this, we need to compare the descendants and fates of the allosaurs and the wood wasps.

For all their grandeur, the allosaurs faded away. They may have been the forebears of the Cretaceous tyrannosaurs and velociraptors, but even these dinosaurs fell into decline, and around 65 million years ago the carnivorous theropods were snuffed out entirely. Such exceptionally large predators just didn't have the staying power to last forever, but more importantly, their ecological niche requirements were massive; terrestrial ecosystems could only support a few such stupendous predators at the same time. There was no way they could possibly diversify into hundreds or thousands of descendant species.

Consider, by way of contrast, the history of the Jurassic wood wasps. Sometime in the Early Jurassic years, a band of rebellious young wood wasps rejected the vegetarian diets of their ancestors and decided to eat beetle larvae. And so the dynasty of the parasitic Hymenoptera was born. Over the passing years these meat-eating wasps diversified and specialized to feed on a variety of multifarious insect species that had already evolved in the forests. By the end of the Jurassic there were hundreds of these parasitic wasp species, by the Late Cretaceous there were thousands, and currently there are hundreds of thousands, possibly millions, of descendant species.[5]

We all have a pretty good intuitive grasp of what predators are: aggressive animals like *Anomalocaris*, scorpions, meganeurid air dragons, fin-backed reptiles, tyrannosaurs, and praying mantises that stalk and eat other (prey) animals. Easy enough to grasp—but there is an easy-to-overlook subtlety here: predators require multiple prey individuals and must keep hunting to survive. This approach worked fine for thousands of species over hundreds of millions of years, but in the Early Jurassic the wood wasps broke the mold and invented an entirely novel predatory behavior when they killed and ate only one other individual animal (called the host), that was large enough to feed them to adulthood.

Although the term "parasitic" is loosely applied to a variety of unrelated insects such as lice, fleas, certain flies, and parasitic wasps, all of which live at the expense of one other animal, the parasitic wasps are different from lice and fleas because they feed extensively on the host animal, eventually killing it. This deadly behavior is so important ecologically that we use another term to describe organisms that do it: parasitoid. A parasitoid is any parasitic organism, more often than not a wasp or fly, which causes its host to die. The Jurassic parasitoids didn't just find a new protein-rich meal, they narrowed their ecological niches to smaller dimensions than those of any previous predatory animals and in doing so allowed their descendants to live in a multitude of previously unoccupied microscopic niches. From that time onward, parasitoids dominated the diversity of terrestrial communities, and by their selective killing behaviors they shaped the richness and abundance of both the insect and plant communities. The scientific fact that the most successful bloodline of parasitoid animals is descended from a clan of log-chewing Jurassic wood wasps is well established. It's broadly supported by independent evidence from the anatomy of living wasps, the ecology of their feeding behaviors, the fossils of Mesozoic wasps, and DNA evidence from modern wasps. But why did wood wasps in particular evolve parasitism?

This is a good point to pick up a discussion thread from the last chapter, where we introduced the xyelid sawflies. Over the Late Triassic and Early Jurassic years, sawflies became among the most successful forest insects by diversifying into myriad vegetarian sawfly clans. As new plant species evolved, sawflies successfully colonized them, and new species coevolved with the forest plant community. The sawflies also subdivided the plants into multiple feeding niches. Some fed externally on leaves, while others learned to conceal themselves by tunneling and mining in leaf tissues or by hiding and feeding in leaf shelters constructed by tying together leaves with threads of silk. As sawflies were adapting to feed on different plant parts, so were other kinds of insects. As a result, competition for all edible plant parts increased, and sawflies responded by further varying their diets. Some specialized by tunneling into plant stems and others ultimately delved into the thicker woody tissues, evolving into the wood wasps. Their success in this endeavor was propelled by the same useful tool that promoted the first sawflies: the female ovipositor.

The Story of the Sting

We tend to think of wasps' ovipositor, from which the stinger, or sting, evolved, as being a simple structure, like a hypodermic needle adapted for injecting eggs. It is actually far more complex. If you were to slice a wasp ovipositor in cross section and examine it microscopically, you would find that it is not one hollow tube but three or four separate interlocking shafts, each of which can be moved independently of the others.[6] This allows the tip of the ovipositor to drill into hard substances, such as plant stems and wood, and be directed with great accuracy. To better understand this idea, try the following exercise. Clasp your hands together by interlocking your fingers and wrapping them tightly around your knuckles. Now extend the first two fingers of each hand, so that exactly four fingers are extended straight. This represents the form of a wasp ovipositor. Look at your fingers end-on and you will see that there is a little hollow space in the middle. This is like the tube through which the wasp egg passes. Now try moving some of your extended fingers, while holding the others in place. For example, try sliding your right fingers downward while keeping your left ones stiff. You will notice that the tip formed by the left fingers is turned to the right, even though they are not being flexed. This is how a female wasp is able to direct the tip of her ovipositor; by retracting the four shafts independently or together, she can move the tip with precision in various directions.

The transport of eggs, hypodermic-fashion, across a microscopically thin tubular pathway required the evolution of microscopic eggs with highly flexible shells, eggs that could be distorted from an oval shape into a long, thin, sausage shape while being forced through this minute tubule. But more importantly, wasp eggs needed a mechanism that would transport them through the ovipositor. In this case, the mechanism was fluid pressure: the ovipositor is a hypodermic needle through which wasp eggs are literally squirted. Consequently, right from the start, wasp ovipositors evolved with a variety of associated liquids. These fluids initially came from female reproductive glands, and provided not only lubrication inside the ovipositor shaft, but also fluid pressure for physically transporting the eggs. They also allowed for the early evolution of wasp venoms.

From sawflies to wood wasps to parasitic wasps, these venoms diversified to accomplish an array of useful functions that we still see today. Some sawflies inject venoms into plant tissues along with eggs, and these venoms induce unusual plant cell development, causing galls to grow. These growths form protective sites for egg development, as well as safe areas and nutritious tissues for larval feeding. In other cases the injected venoms may have antibiotic properties that protect the eggs from the ravages of microbial growth. Among wood wasps, gooey venoms were adapted to promote the growth of fungus. As I mentioned in previous chapters, wood, with its high amounts of nonnutritive lignin and cellulose, is the least digestible part of the plant for an insect. So, like the wood-boring beetles before them, wood wasps adapted to eat the fungus that grows inside decaying wood.

The immature wood wasps were bound to bump into the juicy larvae of various wood-boring beetles from time to time, which were bound to be more delicious and nutritious than either rotting wood or fungi. The jump from chewing on fungi to chewing on the meat of other insects does not seem all that great when one considers that on the tree of earthly life, animals are more closely related to fungi than we are related to plants. In other words, animal tissue is more like the tissue of a mushroom than that of a leafy green plant. Nutritionally and physiologically, the change was not that enormous.

More challenging, perhaps, were the vital behavioral changes involved in switching from a vegetarian to a carnivorous life style. A wood wasp chewing on fungi in a rotting log didn't have to contend with the fungi fighting back. Beetle grubs, however, were not totally defenseless. They could still wiggle around in their tunnels, and when

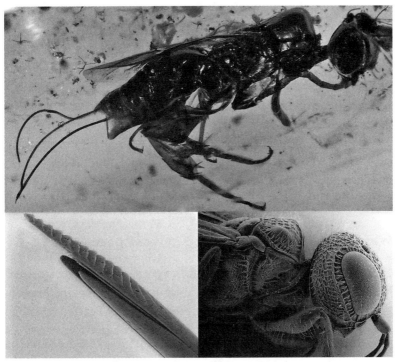

FIGURE 8.2. Megalyrid wasps (order Hymenoptera, family Megalyridae) are among the oldest living examples of parasitoid wasps; the family is thought to have evolved in the Jurassic period. *Top*, a female of an undescribed extinct megalyrid species in mid-Cretaceous amber from Myanmar with a long ovipositor, estimated to be about ninety-nine million years old. (Photo by Vincent Perrichot.) *Bottom*, the head (*right*) and ovipositor (*left*) of a modern, undescribed *Dinapsis* species from Madagascar, a rare surviving example of this ancient wasp group.

attacked they could bite just as well as the wasps. Moreover, a wood wasp's developing eggs would be defenseless against a beetle larva's chewing mouthparts. Clearly the wood wasps needed a secret weapon to tip the scales in their favor.

The adult wasps, not their larvae, developed the trick that gave them the decisive advantage, and once again, the female ovipositor made all the difference. At about the same time that some wood wasp larvae were developing a taste for meat, some of their mothers were refining their egg-laying skills and trying out new kinds of venoms. These mother wasps would drill adroitly into the tunnels where large beetle larvae were feeding, then poke their ovipositors directly into the larvae and inject a new kind of venom that induced permanent paralysis. Then they would extract their ovipositor a bit and carefully place their eggs on the paralyzed beetle. Consider the advantage of venom that did not kill an insect, but rendered it motionless. If a beetle larva were killed by an adult wasp, it would begin to rot and decay, and the decomposition process would endanger the developing wasp egg. Paralyzing the host insect with venom instead is a cheap and efficient way to preserve the meat until the egg can hatch and the baby wasp can safely begin feeding. It's pretty much like setting a full food dish next to a lazy dog: the wasp larva doesn't have much to do other than sit there and chew on a big chunk of fresh meat.

Which Way to Eat an Oreo: Two Kinds of Parasitism

The sort of external parasitism that I've been describing has existed for about 150 million years. Termed "ectoparasitism," which literally means "feeding as a parasite from the outside," it became, over the intervening years, an important factor in the evolution and success of the wasps in particular. But the origin of parasitism is really just the beginning of a much bigger story. Although the first parasitoid species were very successful, diversifying and chewing on whatever young beetles they could find in the decaying wood for millions of years, they stayed restricted to that particular habitat until some entrepreneurial wasp came along with another new approach. Sometime in the Late Jurassic or Early Cretaceous the parasitoid wasps figured out how to feed inside other animals as an internal parasite. This more refined method of parasitism is what we term endoparasitism. It literally means "feeding as a parasite from the inside," and looking at the modern insect world, the vast majority of parasitic species are of this second sort.

Although ectoparasitoids' early success in the Middle Jurassic was mainly due to their strategy of feeding externally on only one small organism, this approach had a major drawback. The host insect was permanently paralyzed and the immature wasp needed ample time to hatch from its egg, then devour the entire beetle grub. The process took many weeks, at least, and could work only in concealment. If exposed, the young wasp would be revealed to predators and subject to harsher environmental factors: desiccating sunlight, temperature extremes, wind, and storms. Therefore external parasitism was mostly limited to protected microhabitats inside plant tissue, which hindered what ectoparasitoids could ultimately achieve. Now, as then, the great majority of external-feeding parasitoids are associated with immature insects found inside plant tissues. The internal feeders, in contrast, were unfettered by these habitat constraints. The host became the endoparasitic wasp's niche and its entire habitat during its youthful existence (endoparasitism was literally the discovery of niches within other insects). Once inside the host, the endoparasitic wasp larva became entirely portable, and it could exist in any habitat where the host might exist. As a result, endoparasitic wasps were able to diversify and feed inside virtually all kinds of Mesozoic insects. A Pandora's box of feeding behaviors was opened for the wasps.

Once again, an important event in the history of wasps was the refinement of egg-laying behavior and the female ovipositor. Some females quit laying their eggs on the outside of the host and started injecting them directly inside it. Stated succinctly like that, the transition to endoparasitism sounds simple, but it wasn't. Remember that female wasps were already drilling and stinging host insects for millions of years, injecting paralyzing venoms with their ovipositors. They could have easily inserted their eggs inside other insects along with venom as soon as they developed parasitic behavior. But they didn't. Instead, they pulled out their ovipositors from inside the hosts and laid their eggs on the outside, allowing their young wasp larvae to feed externally. The reason they didn't initially place their eggs on the inside is that being there is a lot more challenging than being on the outside. Insects have an open circulatory system, so their inside is a sack full of organs bathed in a pool of blood. That blood, like our own, contains cells that defend insects against microscopic invasion. Early parasitic wasp venoms may have paralyzed a host's muscular system, but they did not incapacitate the immune response of its blood cells. A small egg placed inside an insect's body cavity would be swarmed and encapsulated by these cells and killed. Successful endoparasitism required that wasps evolve an array of special adaptations to the internal environment.

The first step to successful internal parasitism was yet another refinement in precision egg laying: some wasps laid eggs directly into the host's nerve or muscle tissue, thereby avoiding its blood—and immune system—entirely. This is a nice adaptation, as far as it goes, but sooner or later the egg needs to hatch and the young larva must move around and eat. It's hard to avoid the blood entirely, and in fact, there is a very good reason to want to be there: insect blood is a pool of nutrient-rich fluids. So ultimately the most successful internal parasitoids were the ones that invented ways to compromise the host's immune system. Once again the mother wasps helped their offspring. Along with eggs, they injected venoms, some of which were modified to help disable the immune system—but somewhere along the line an even more unexpected event occurred: a symbiotic relationship was forged between certain viruses and the wasps.

We all have heard how dirty hypodermic needles can transfer viruses. Back in the very early days of internal parasitism, one of the wasps managed to soil its own hypodermic ovipositor with some virus particles. This happened fortuitously, but then those particles were injected, along with a wasp egg, into a hapless host insect. The virus replicated itself within the host, disabling its immune system but not harming the wasp larva. At the same time, the virus was able to imbed itself inside the developing wasp's body and so was able to escape and eventually find its way to another potential host. It was a win-win situation for both the virus and the wasp.[7]

Once host immune systems were disabled, wasp eggs and larvae could wallow safely in insect blood. An external parasite's egg has a tough outer shell, which protects it from environmental factors, and a large protein-rich yolk, which feeds the developing embryo until it hatches into a larva. An egg placed in blood, on the other hand, does not have such a thick outer shell; it floats in a protein-rich liquid environment and, with a thin shell, absorbs nutrients directly from the host's blood. Now the endoparasitic wasps only needed a mechanism for extracting nutrients, so that their eggs could survive with little, if any, yolk. If less yolk were needed, then females could produce eggs more easily and lay more of them. And so the endoparasitic wasps evolved a structure called a trophamnion, which works like a parasitic placenta. The trophamnion consists of a cluster of cells, closely connected to the embryo, that absorbs and transfers nutrients directly from the host's blood. These cells feed the embryo it until it develops into a larva, which bursts from its egg and swims away. The trophamnion's benefits do not end there. As the egg hatches, its cells disassociate and move independently into the host's blood pool. These former trophamnion cells continue to extract nutrients from the insect's blood and work to further disable its immune system. As they continue to feed, they grow and eventually morph into giant cells, called teratocytes, which the young wasp larva also consumes as it swims about and dines on the host's blood and tissues.

Wasp larvae adapted to life in their miniature aquatic environment by developing the ability to swim with long, taillike appendages, which they whip back and forth, and the ability to breathe with closed, gas-filled tracheal systems, which, since they have no open breathing holes, prevent water from flooding into them. The larvae also developed a thin cuticle without much hard skeletal material along their body wall. This allowed them to breathe directly through their body by a process known as cuticular respiration, just like many other aquatic insects that live in ponds and streams. Indeed, the wasps are actually the tiniest of aquatic insects and also the most diversified group of aquatic organisms.

Although a larger aquatic insect living in a pond has a lot of room to move about, a parasitic wasp larva swimming in insect blood is limited to a small enclosed space. It literally lives inside a bowl of soup, its only food. This presents a special problem: anything that eats and grows must also produce waste, so how does the larva continue to develop without fouling its food and living environment? We can appreciate the importance of teaching kids not to pee and poop in the swimming pool and bath tub—all the more important when the pool is a food source as well. Young wasps solve this problem by simply accumulating waste inside their bodies and never defecating. They assimilate nutrients very efficiently with the middle part of the digestive system, but the hind part is closed, forcing waste into the rear end. When a wasp larva is done feeding, growing, and molting through several stages, it exits the host to spin its own silk cocoon, pupate, and finally transform into an adult. Upon emerging from its cocoon, the full-grown wasp voids its larval waste for the first and last time.

Perhaps the most curious thing about internal parasitic wasp larvae is their startling array of body forms.[8] The demands of life inside another insect are different as the larva grows, so with each molt it has an opportunity to acquire a new form more perfectly adapted to the needs at a particular stage of life. One might guess that, after hatch-

FIGURE 8.3. The caterpillar of a papilionid butterfly from Costa Rica, which has been parasitized by an endoparasitoid braconid wasp, *Meteorus papiliovorus*. The larva of the parasitic wasp has recently consumed most of this caterpillar internally, chewed an exit hole (visible on the right), and spun its silk cocoon suspended by a short thread. (Photo by Kenji Nishida.)

ing from its egg, eating would be its main concern. Finding food is easy enough, however; the larva effortlessly imbibes its liquid meal. It has a bigger worry: sheer survival. The host's immune system may be disabled, but another source of immediate danger might be present: competition from other parasitic larvae, either of another or the same species, residing inside that same host insect. Because these competitors might pose a mortal threat, young wasp larvae have become hyperspecialized for defense. Many are born with a large head, powerful sickle-like jaws, and as I already mentioned, a long, taillike appendage for rapid locomotion. Their first job in life outside the egg is to swim about and aggressively eliminate their rivals.

For the middle part of its life, the wasp larva exists as a featureless white maggot. It has eliminated its competition, so it no longer has large defensive jaws. Given that its food consists of liquids, small particles, and cells, and that it does not even need to chew, its head becomes greatly reduced. The larva doesn't have to travel, so it loses its swimming appendage but does not grow legs. Because its cuticle is thin, gas can exchange across its body surface, and the larva can stretch, making rapid feeding and growth easy. The wasp larva keeps this simple body form for one or more molts, over the middle part of its life, when it has little to do other than float about, suck in food, and get bigger. Its hind digestive tract remains closed.

As it gains the large body mass needed to transform into an adult, the wasp larva prepares to pupate and exit its host, whose inside is largely devoured and destroyed, and will soon begin to decay. For most larvae, this is now a good time to get out. During the last molt several major changes occur. The mature larva acquires open spiracles, and becomes capable of breathing gaseous air through tracheal respiration. It develops a more complete head with mandibles capable of chewing an exit hole, and spinnerettes capable of building a silk cocoon. Silk glands suddenly and rapidly develop.[9] Last but perhaps not least, the mature larva grows a complete digestive tract, and is finally able to eliminate its stored body waste.

Double Vision: Two Other Ways of Looking at Parasitism

The evolution of endoparasitism led to an exceptional burst of species originations, as various wasps diverged and adapted to life inside other insects. But the wasps still had one important trick left, which sparked yet another explosive radiation of species. You will recall that parasitism began when wasps attacked large host insects in wood, insects paralyzed by female venoms. Parasitism of this sort, where the host is permanently paralyzed, is termed idiobiosis, and the wasps that display this behavior are called idiobionts. But the vast majority of modern wasps don't do this anymore. The only real benefit to paralysis was in preventing the host from harming a young wasp. Once endoparasitism evolved, however, paralysis was not only not required, it may have become a real drawback. Consequently, the vast majority of endoparasitic wasp species have developed another behavior where the host insect is either not paralyzed at all, or is only temporarily paralyzed during the egg-laying process. Parasitism of this sort is termed koinobiosis, and the wasps are called koinobionts.

At first blush, the difference between idiobiosis and koinobiosis may seem rather small, but the implications for wasp evolution were enormous. The key distinction is that while the idiobionts turn the host into a defenseless hunk of preserved meat, the koinobionts insidiously allow the parasitized insect to live on, continue feeding, grow, and even molt, after they have placed their parasitic egg. This allows the wasps to attack a greater diversity of life stages. While an idiobiont is restricted to hunting the biggest host it can find because its larva can feed only on the meat provided, a koinobiont has the option of attacking much smaller hosts. Many koinobiont species put their eggs into very young insects, which are always more numerous than older ones, at a time when the host is so small that it cannot provide the parasitoid offspring with enough food to grow into adulthood. But a koinobiont larva gets around this problem by developing slowly or delaying its own growth; the host insect is allowed to live until it attains enough biomass to properly feed the larva. By delaying development just a bit longer, a koinobiont wasp allows its host to begin pupating, which often involves seeking a safe hiding place, and perhaps forming a silk cocoon. The wasp is therefore able to take advantage of the host's protective behavior.

Idiobiosis is a great strategy for attacking concealed insects, but it's a poor strategy for attacking exposed ones. If an exposed insect is permanently paralyzed, it easily becomes a target for scavengers and predators, and the young wasp is killed along with its host. So perhaps the greatest benefit of koinobiosis was that it enabled parasitic wasps to escape from dead wood and assail hosts which lived and fed externally on plants. With one fell swoop, koinobiosis broke the bonds of habitat confinement in the late Jurassic forests.

Secretive Societies with an Anal Fixation

While the wasps became supremely successful by inventing ways to escape the confinement of rotting forest logs, another emerging clan of Jurassic insects thrived by evolving a whole new way of living inside them. These entrepreneurial insects were the first termites, also known as isopterans, and they developed complex group-living behavior, making them the first truly social insects. When entomologists speak of social organisms, we don't just mean that they gather together in groups. Way back in the Devonian and Carboniferous years, hexapods and insects aggregated—creatures such as springtails, bristle-

tails, mayflies, and roaches may have lived together in large numbers—but we do not define them as social. Strictly speaking, social
insects do reside in groups, but they also have three broad characteristics.

The first requirement of social behavior is a longer adult lifetime,
such that two or more generations of individuals coexist. In most
nonsocial insects, the adult lays eggs and then dies; most parents do
not live to see their children mature to adulthood. The social insects'
longer life span allows them to accomplish the second requirement
of sociality: cooperatively caring for the young. Social insects provide
the next generation with food, remove their waste, and protect them
from predators and parasites until they can successfully mature and
contribute to the colony labor themselves. Last but certainly not least,
social insects show a division of colony labor; this has allowed specialized forms (castes), which perform particular roles in the colony, to
evolve. The vast majority are sterile workers, individuals that do not
produce their own children but instead help raise their mother's offspring. These workers construct the nest, forage for food, and feed the
growing young. They do not defend the colony, however; this special
task is performed by the soldier caste, sterile individuals with massive heads and mouthparts so specialized for defense that they can't
feed themselves and must be nursed by the workers. Only a very few
individuals in a termite colony actually have their own children: the
reproductive kings and queens, the first royalty in the history of the
planet. Once the king and queen have established a new colony and
raised the first generation of workers, they just sit back and enjoy the
fruits of their labor.

Termites are often regarded as social cockroaches. The most primitive living species are indeed similar to the wood-eating cockroaches
(Cryptocercidae), and it is generally agreed that termites evolved
from roachlike ancestors.[10] The key to termite behavior and existence is their ability to digest cellulose from woody plants. Like their
near cousins the wood roaches, they accomplish this difficult feat by
housing symbiotic microorganisms in their digestive tracts. Like all
other insects, they have an external skeleton, and their foregut and
hindgut are lined with skeletal material. Therefore, when they periodically molt their skeletons, termites lose their symbionts as well,
and they must acquire new ones or else they will starve to death. They

FIGURE 8.4. Considered the most phylogenetically primitive of living termites, *Mastotermes darwiniensis*, from tropical northern Australia, is the only living termite species that lays eggs in pods, like cockroaches. Pictured are three social castes: a soldier (*upper right*), a dark-colored neotenic reproductive individual, and several pale-colored workers. (Photo © Barbara Thorne.)

get their symbiotic gut microorganisms by a process called anal trophallaxis—literally by eating the feces of other termites. As repulsive as this behavior may sound, it is crucial to their survival, and without it some of the world's most impressive and influential societies might never have evolved.[11]

Anal trophallaxis, ironically, solves another serious problem of subsisting in large societies: sewage removal. When insects (or other animals) live together in groups, their accumulating feces can promote the spread of microbial pathogens, such as bacteria and fungi. It can also attract predators. When caterpillars feed in groups, for example, they are far more likely to suffer from diseases or be preyed upon than solitary caterpillars dispersed in the forest. The termites avoid all this not only by eating their own feces but also by using it to build tunnels and arches within their nests.

The earliest termite families probably contained only dozens or hundreds of individuals, but they were so successful that from them emerged over 2,900 modern species, whose colonies can include upward of several million individuals. Each species has its own unique habits and life style, but each contributes powerfully to nutrient cycling and vertebrate food webs. Because of their diversity and abundance, termites are among the chief decomposers of wood and other plant materials, especially in the humid tropics. They help create and move soils, and they are voraciously consumed by other animals. Consider that a single towering mound of an African *Macrotermes* can be 20 feet tall and contain more than 2 million worker termites. Compared to the termites' own body size, if scaled to human proportions, one of these mounds is taller than any skyscraper. Along with the mounds of other termite species, they are all the more impressive when you consider that they are single-family dwellings. A *Macrotermes* queen can live for 10 years, sometimes laying as many as 30,000 eggs in a single day; over her life she may lay as many as 100 million. How's that for a family picnic?

Of Lice and Hen

We can't go picnicking in Jurassic park without mentioning the feathered dinosaurs. Much has already been written about one of the oldest-known birds, *Archaeopteryx*, which had a scaly dinosaurish head, large feathered wings with claws, and a long feathered tail.[12] How did *Archaeopteryx*, along with other early birds, learn to fly? One popular idea is the arboreal hypothesis of gliding flight, which suggests that birds evolved from small, feathered, tree-dwelling dinosaurs that first flew by gliding from tree to tree. This is easy enough to accept, as it simply proposes that the birds mimicked a successful strategy pioneered by the insects about 150 million years earlier. But the gliding-bird hypothesis has a couple of drawbacks. Why did the ancestral ground-dwelling dinosaurs bother to move up into the trees? And what factors might have caused them to develop large feathered front limbs, even before those appendages were capable of flight?

The cursorial hypothesis gets around these problems by proposing that birds learned how to fly from the ground up. Although hotly debated, this idea remains close to my heart for a simple reason: it asserts the fundamental importance of insect feeding as a driving mechanism in bird evolution. The cursorial hypothesis suggests that birds started out by running along the ground chasing insects. It further suggests that these animals first evolved large feathers on the forelimbs not for flight, but as improved insect-catching devices. The idea of ancient feathery dinosaurs rapidly running along through the forests, using their front legs as fly swatters, is not far-fetched. We have already come to grips with the notion that many small ground-dwelling Triassic dinosaurs were omnivorous carnivores and therefore highly insectivorous. And birds clearly evolved from these very same dinosaurs, little ones, like *Ornitholestes*, that would most likely have fed extensively upon bugs. It makes perfect sense to suppose that the earliest protobirds would also have eaten insects, and that strong selection would have favored any behaviors that improved their ability to rapidly catch insect prey. The main drawback to this hypothesis is evidence suggesting that *Archaeopteryx* spent little or no time on the ground: their claws show very little wear. But the two ideas need not be mutually exclusive. While *Archaeopteryx* may have been a flier living in the treetops, its immediate ancestors might have been ground-dwelling insect chasers.[13]

Whatever you think of this wing-swatter idea, the history of the birds' emergence is intertwined and codependent with the insects' history and success. The first birds could move into the air, and what better reason than to gobble up the highly nutritious swarms that were everywhere around them? What the birds lacked in size, they made up in numbers, evolving into more than eight thousand modern species, the majority of them insectivorous.[14]

One might suppose that once the feathery dinosaurs burst into the skies, they spelled trouble for the insects and hunted some of them to extinction; however, there's not much evidence to support this. One group did disappear about the same time that the early birds arose—the titan insects—but they were overly large and noisy. They might have been all too easily hunted. Other groups appear not to have been impacted and actually became more diverse. We can safely assume that, from the first, birds were powerful selective forces which drove insect evolution in several ways. Day-active feeders, they probably influenced the evolution of many insects' night activity. And as visually searching predators, they helped stimulate the evolution of crypsis, aposematic coloration, and mimicry. So when dinosaurs finally feathered their way into the air, they didn't hinder insect success. Bugs had been flying for 150 million years by the time birds finally came along; by then they were pretty good at it.

As birds diversified, insects were probably quick to exploit them. A new group of parasitic insects, the biting and chewing lice, were likely early bird colonists, though they're not to be confused with their more common cousins, the blood-sucking mammal lice, which evolved much later. Both the biting and the blood-sucking lice (order Phthiraptera) originated first as chewing lice that themselves probably evolved from bark lice feeding in the nests of early birds. Since the chewing lice have such a strong affinity for living among feathers—so strong that we often call them bird lice—we can't rule out the possibility that even earlier, ground-dwelling, feathered dinosaurs might have already had lice prior to the birds' emergence. However they came to be there, the bird lice grew very successful in that miniature feathery forest: today, there are more than 1,200 modern species.[15] The potential fauna of parasitic lice was limited by the available vertebrate hosts; they could never attain the levels of hyperdiversity displayed by the wasps, which exploited the extreme species richness of other insects. Despite this constraint, the origin of the parasitic lice nevertheless remains another page in the saga of how insects successfully colonized our planet.

During the Jurassic years, Laurasia and Gondwana continued to drift apart, creating more coastlines, changing continental climates from dry to wet, and transforming arid Triassic deserts into moist forests, whose towering redwood-like conifers soared above even the tallest *Diplodocus*. Although a multitude of new dinosaur species stalked through fern meadows and clamored among ginkgo branches in the real Jurassic Park, most of the period's diversification occurred at a microscopic level, among the bugs, beetles, wasps, and flies. As the Jurassic drew to a close, legions of parasitoid wasps drilled into dead trees for beetle larvae, while termites chewed delicately amidst the ground litter, and tenacious lice nestled among the feathers of small dinosaurs. But insect and plant evolution in the Mesozoic forests was about to become vastly more visible. During the dawn of the Cretaceous, a profusion of colors erupted as the forests bloomed with the hues and scents of the first flowers. The Cretaceous period also saw the first bees, flower flies, and butterflies. Nearly three hundred million years had passed since plants colonized the land. Why did the world take so long to blossom?

9 Cretaceous Bloom and Doom

The evolution of the flower in all its complexity of form, color, and scent has gone hand in hand with the evolution of pollinating insects.

V. H. HEYWOOD, *Flowering Plants of the World*

Perfectly reasonable scientists, who pride themselves on their caution when dealing with their own specialty, indulge on the wildest flights of fancy when it comes to cracking the mystery of the Cretaceous killer.

ROBERT T. BAKKER, *The Dinosaur Heresies*

The king is gone, but he's not forgotten.

NEIL YOUNG, "My My, Hey Hey"

As I write these lines it is a frigid February in Wyoming. The landscape is about as cold and barren as can possibly be, and yet I am thinking of flowers. Valentine's Day will not let me forget them. The television networks, radio stations, newspapers, and Internet all remind us to buy flowers for the ones we love, and for a few short days, in the bleakest of winter, the stores are saturated with bundles of roses and a lush variety of other blossoms in dozens of colors.

Unless we have allergies, flowers generally make us feel good. We plant them in our yards and gardens around our homes and workplaces. We culture them in small containers and bring them indoors. We like their riotous colors and the extravagant forms of their petals, and so we create floral art, drapery, and wallpaper; grace our fabrics with floral images; and swaddle ourselves in flowery clothing and jewelry. Sometimes we even tattoo flowers onto ourselves. We especially like their smell. We harvest or duplicate their scents for aromatherapy and incorporate them into perfumes, soaps, oils, lotions, shampoos, deodorants, and candles.

Perhaps our attraction to flowers might be another example of the phenomenon that Edward O. Wilson has termed biophilia, the innate love that humans have for nature and for other living things. Because we have evolved over millions of years in nature, maybe we possess a genetically programmed yearning to surround ourselves with its more pleasant aspects—flowers being one of its best-loved items. But we seem to like flowers more than we do other living creatures. We certainly don't feel quite the same way about fungi, salamanders, frogs, spiders, or snakes, all of which can be colorful and interesting to look at. Some stink bugs have very pleasant odors, nicer indeed than some flowers, but we don't give them to our wives or girlfriends on Valentine's Day. Perhaps at a very basic level flowers attract us with the same characteristics that lure insects to them. Their vivid colors and curious shapes allow us to spot them from a long distance, just like a bee. At close range we react positively to their odors and sweet nectars, just like a butterfly. Flowers, bees, and butterflies are so commonplace today, it is easy to forget that the earth was not always filled with them.

Rift, Shift, and Dinosaurs Adrift

We tend to hear more about the Cretaceous's catastrophic finale and not much about its long years. The period ended spectacularly with the colossal asteroid impact that (presumably) extinguished the *Tyrannosaurus rex* and its kind, and that (finally) allowed mammals to emerge from the carnivorous dinosaurs' long shadow. We will hear more about that later, but for now let's focus on what things were like during the Cretaceous, which lasted about seventy-nine million years—a long time, even by cosmological standards.

With the end of the Jurassic and the onset of the Early Cretaceous, about 130 to 145 million years ago, the composition of this planet's big animal communities noticeably changed. The *Apatosaurus, Diplodocus*, and their relatives faded to extinction, and the Cretaceous forests became filled with new communities of large herbivores, most notably the duck-billed hadrosaurs and the horned and frilled *Triceratops* and its multipronged relatives. Meanwhile, the southern supercontinent of Gondwana was violently fragmented, wrenching apart startled dinosaur herds. The land mass now known to us as South America began to split away from what is now Africa. With each massive earthquake, rifts formed from the south and north and ocean water rushed in, until eventually these areas were completely separated by a water gap and a narrow, infant South Atlantic Ocean was formed. The South Atlantic sea floor has continued to grow and widen since then, gradually but relentlessly pushing South America and Africa further and further apart, but it was in the Early Cretaceous that South America became a large island continent, at first just narrowly separated from Africa and already widely separated from North America (the Central American land bridge did not join South and North America together until just a few million years ago).

If we were to view the Cretaceous continents from outer space, we could recognize at least the shapes of modern areas like South America, Africa, India, Antarctica, and Australia, but their positions were stunningly different than they are today. South America and Africa were separated by a narrow waterway that resembled a winding channel rather than a modern ocean; Antarctica was located further to the north, much closer to South America and Africa; and Australia, tilted on its side, was still very close to Antarctica, divided by only a thin gap. In the Early Cretaceous these southern continents had mild climates, and animals and plants were dispersed along the southern corridor from South America, across the northern coast of Antarctica, to Australia in the east. South America was not the Cretaceous's only distinct island continent. The subcontinent of India had already fragmented from the eastern coast of Africa, and over the period a fully isolated India drifted north and east across what we now call the Indian Ocean.

Although the shapes of Cretaceous India and South America might have looked familiar, their terrains would have been unrecognizable. Most notably, they were relatively flat; the modern mountain ranges that we know as the Andes and the Himalayas had not yet formed.[1] Moreover, in the Early Cretaceous the Amazon and Ganges rivers did not exist. More ancient rivers drained across the lowland forests as duck-billed dinosaurs browsed in the misty twilight. Yet as strange as this scene might seem to us now, it also had a growing familiarity: in the Cretaceous forests the flowering plants first evolved and rapidly proliferated, and myriad flower-associated insect communities developed.

Dance of the Sugar Plum Fairies:
The Coevolution of Insects and Flowers

The flowering plants, the blossom and fruit-producing organisms known to botanists as angiosperms, may have first evolved in the Jurassic period or earlier, but they were initially rare woody shrubs restricted to wet forest habitats. We have fossil flower pollen dating to the Early Cretaceous, 134 million years ago, and fossil leaves and flowers dating to 124 million years ago, and we know that by 120 million years ago the first angiosperms, including such recognizable species as water lilies and magnolias, quickly radiated and diversified. By the Middle Cretaceous (and on to the present day) angiosperms had become the dominant plant species.

The early angiosperms' method of reproduction is similar to that of the Carboniferous Gnetales, which predated the flowering plants by at least 160 million years and resembled the coniferous cycads, primitive seed plants with stout woody trunks, topped by crowns of stiff evergreen leaves, that had dominated the dinosaur-ridden forests since the Triassic. The Gnetales used pollen to reproduce and had two forms: some had pollen-producing structures and others pollen-collecting structures.[2] Their female reproductive parts don't look like anything we would call a flower, but they function in the same way—so although the gnetaleans are not classified as angiosperms, they might reasonably be considered the most ancient flowers. More importantly, it appears that the gnetaleans were insect-pollinated. Modern species, such as *Ephedra antisyphilitica*, are known to produce sticky droplets of pollination fluid at the tip of their flowering structure. This fluid grabs microscopic pollen grains but also attracts insect visitors with its sweetness.

Sweet nectar and nutritious pollen allowed flowers—and insects—to overrun the planet. Plants produce them in sacrificial abundance, enough to feed ravenous hordes of flies, beetles, wasps, and moths, all of whom, in turn, scatter the protein-packed pollen that sticks to their hairs. The nutritional benefit to the insects is obvious, but how did the plants profit from this mutualism? Plants are rooted in place; they cannot get up and trot away in search of mates. Until the Cretaceous, their distribution was limited mostly by the constraints of wind pollination. But with the insects' assistance—and thanks to the energetics of insect flight—plants at this time could spread their genetic material over long distances. Now they could exist as widely dispersed populations, scattered in forests with little wind movement.

The benefit of pollination systems to both plants and insects is clearly imprinted in the fossil record since the Middle Cretaceous. Fossils from this time are rich not just with diverse, new flowering plant species but with new species of flower-associated insects. Of particular note is the appearance and diversification of new flies with long, hollow, tubelike mouths modified for dipping into and sucking nectar from deep inside flowers. These long-beaked flies—the early bee flies, flower-loving flies, and tangle-winged flies (families Bombyliidae, Mydidae, and Nemestrinidae, respectively)—are among the earliest of the highly specialized flower-feeding insects. Moreover, by looking at remnant living Gnetales species such as *Ephydra*, as well as still-living primitive flowering plants like water lilies and magnolias, we can see that these ancient kinds of flowers achieved their success by attracting communities of assorted generalized insects—all with a taste for flowers. Close codependencies, like those between orchids and certain bee species, are the result of coevolution over the past hundred million years or more.

The key groups of plant-eating insects—stick insects, katydids, plant bugs, leafhoppers, thrips, beetles, sawflies, flies, and moths—were already in place before the Cretaceous began. All enjoyed intense bursts of speciation that paralleled the plants' rapid diversification. But of all these evolutionary explosions, none can rival the rise of the moths and butterflies during the Late Cretaceous. No other animal group since has more successfully colonized the flowering plants. What propelled them to evolutionary greatness was the feeding habits of their immature larval stages—caterpillars.

Lepidoptera Domine

Like the flowering plants they chew upon, leaf-feeding caterpillars are so common that it is hard to imagine earth without them. These externally feeding insects burst upon the Cretaceous scene only a mere ninety million years ago, along with the flowering plants' early radiation. They evolved from more ancient, microscopic larvae that tunneled their way through plant tissues, which protected these larvae from predators, wind, and the sun's drying effects. External-feeding caterpillars coped with their harsher environment by evolving thicker cuticles to prevent water loss, but a more serious challenge of life on the outside was simply staying on the plant. Unlike a leaf-tunneling insect, a caterpillar needs to cope with motion. Plants may not get up and run around like animals, but their leaves frequently rustle in the wind and their branches often sway in storms. Holding onto the plant is really crucial. If a caterpillar falls off, it will likely die before it can find the plant again, or it will waste great amounts of energy trying to regain its former position.

The Lepidoptera in particular had a wonderful preadaptation for success in the outer world: silk.[3] While most of us will recall the value of silk cocoons in protecting the transforming moth pupa, we tend to forget the importance of this material to the feeding caterpillar, for which it is a matter of utmost survival. That silk strand is literally a lifeline in the exposed world. As a caterpillar moves about a plant, it is constantly spinning a sticky silk thread, which adheres to the surface. If a caterpillar falls, it will release more silk, with which it can safely descend to another part of the plant, or up which it can return to its starting point. Upon the tips of its abdominal legs are microscopic spines and hooks, called crochets, which grab onto the silk threads while it walks, and keep the caterpillar firmly attached to the plant, even on windy days. The original inventors—and the most successful users—of Velcro, caterpillars also tie together leafs with silk and employ the bundles as feeding shelters to avoid the harsher outer environment and hide from predators and parasitoids.

As the Cretaceous progressed, moth caterpillars and the very first butterflies became increasingly efficient at chewing on plant parts. If there was an edible portion, it seems that caterpillars managed to find it. They eat leaves whole, scrape leaf surfaces, and skeletonize leaf veins, and they also eat buds, flowers, fruits, seeds, stems, and even roots. All of this nibbling and crunching may sound like a total catastrophe for the early flowering plants, but they did not take the assault lightly. As insect feeding increased, the plants responded by evolving more efficient defenses. Some developed thicker cuticles or microscopic spines that are difficult to digest, others thick or gummy saps that bind to insect mouthparts and are impossible to chew. Many other plants evolved chemical defenses that make their leaves bitter, or even

FIGURE 9.1. This well-preserved fossil butterfly, *Prodryas persephone* (order Lepidoptera, family Nymphalidae), is from Eocene-age rocks of Florissant, Colorado, estimated to be at least thirty-four million years old. Butterflies first evolved during the Cretaceous but survived the end-Cretaceous extinctions to become important herbivores in the modern world. (Photo by Frank Carpenter. Museum of Comparative Zoology, Harvard University. © President and Fellows of Harvard College.)

toxic.[4] More than a hundred thousand such defensive compounds have been recognized so far, and new ones are constantly being discovered with the exploration of tropical plants. They seem to rival the diversity of the plant-feeding insect armies and include tannins, alkaloids, cyanogenic glycosides, coumarins, flavonoids, steroids, and terpenoids, to name only a few. While these names may seem foreign, the alkaloid compounds, for instance, include caffeine, nicotine, morphine, atropine, cocaine, strychnine, quinine, and curare. When we sip coffee or smoke a cigar, how many of us pause to reflect on the hundred million years of insect–plant coevolution that made these things possible?

The angiosperms' chemical arsenals did not doom the plant-feeding insects. If anything, they stimulated the evolution of even more creative feeding strategies. If an angiosperm evolves a totally new defensive chemical, this may prevent many generalist herbivores from continuing to eat that particular species. But there are always bound to be a few specialist insects that, because of their unique abilities, discover ways to keep eating the plant. Some insects simply avoid the chemicals. For example, a chewing insect might cut the large leaf veins, thus preventing defensive compounds from flowing into the area where it is feeding, or it might feed selectively on new growth with fewer toxins. Or an insect with piercing mouthparts might feed selectively by puncturing areas of the plant which lack the chemicals. Many insects evolve mechanisms for detoxifying plant poisons or, even better, find ways to incorporate them into their own body's metabolism—thereby turning plant chemicals into their own defense against predators.

Attack of the Bee Girls

The Cretaceous surge of flowering plants and their associated insects also stimulated a parallel explosion of parasitic and predatory groups, since every new insect species that evolved to exploit a flowering plant was itself potential food for insect-eating species. The stinging wasps (Aculeata), for instance, descended from a group of Late Jurassic parasitic wasps that shifted their egg-laying duct from the ovipositor to just above the base of the ovipositor, which was itself modified into a hypodermic syringe used only to inject venoms. This allowed the sting to specialize as a venom-delivery device and ultimately as a purely defensive weapon—a transformation that in the Late Jurassic quickened the conquest of a new wasp group, the nest-provisioning wasps, and in the Cretaceous gave rise to several other new social insect groups: social wasps, bees, and ants.

Probably not much changed for the Jurassic wasps with the new egg-laying duct. The females continued to sting and paralyze insect hosts and to lay eggs on them. However, once the sting evolved, females no longer had the option of drilling and injecting eggs deep into their hosts. They had to lay their eggs directly on easily exposed insects. But the same exposure that allowed the female wasps to easily find these insects also allowed other predators and parasites to discover them, which is why some females started moving their paralyzed hosts to a more secluded spot. With this move, the first nest-provisioning solitary wasps made their debut in the Late Jurassic.

More often than not, a female nest-provisioning solitary wasp will dig a hole or tunnel in the ground or find a hollow plant stem. Then

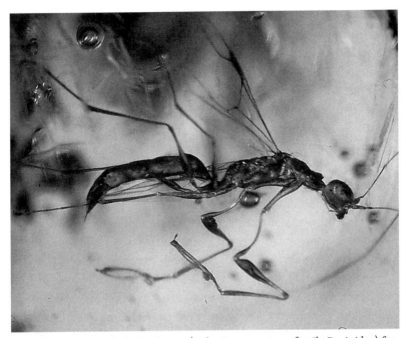

FIGURE 9.2. A female dryinid wasp (order Hymenoptera, family Dryinidae) fossilized in Dominican amber, estimated to be twenty to thirty million years old, with a visible sting. Modern dryinids are all ectoparasitoids of Homoptera adults or nymphs, which they catch with their chelate forelegs and sting to cause paralysis; then they lay a single egg between the host insect's overlapping thoracic or abdominal segments. (Photo © George Poinar Jr.)

she goes hunting. She finds a suitable food item, such as a caterpillar, stings it, and injects her paralyzing venom, just as her parasitoid wasp ancestors did for millions of years. Next she grasps the paralyzed insect with her mandibles or legs. Maybe she flies away with it, or maybe she just drags it along the ground. In any case, with determination and hard work, she takes that insect back to her nest, stuffs it into the hole, and lays a single egg upon it. Then she deftly seals up the nest entrance, and starts the process over again in another location.

As female wasps became better at searching for exposed insects during the Early Cretaceous, they refined their stinger into the supreme hunting tool, and many new nest-provisioning species evolved. As their hunting skills diversified, wasps also grew more creative at making nests. They began to tunnel into wood, hollow out the pithy centers of plant stems, and sculpt nests on vertical rock and cliff faces out of clay or mud. To further hide their nests, some modern wasps close it, fly over it, and scatter sand to erase all traces of the entrance. Other mother wasps pick up a small stone with their jaws and tamp down the soil over the nest entrance. Thus, wasps were the first stone tool users, probably tens of millions of years before any primate or human ever picked up a rock.

High Society

The invention of nest provisioning once again set the stage for the transition from solitary to social behavior. During the Cretaceous period the three major lines of social Hymenoptera all evolved from nest provisioning ancestors. These are the groups known to us as the bees, the paper wasps (including hornets), and the ants.

Bees are, simply, very hairy social wasps that have evolved into plant-feeders rather than carnivores. Primitive bees lay eggs in tunnels in the ground just like many nest-provisioning solitary wasps, and their method of provisioning cells is similar.[5] They just pack the chamber with pollen instead of paralyzed prey. As bee societies further diversified some evolved arboreal nest-making habits and began occupying pithy or hollow tree branch stems. Eventually some were able to sculpt cells and nests from pieces of wax secreted from wax glands, allowing them to occupy larger cavities in trees or cliffs.

Paper wasps evolved the capacity to produce tough paper by mixing macerated wood pulp with their own saliva. They sculpt cells from this "paper," into which they stockpile paralyzed and chewed food items, usually other kinds of insects, for their larvae. Paper wasps have become expert architects, shaping an assortment of sophisticated multicelled and multitiered enclosed shelters; their unique ability to make nest paper and attach it to almost anything has allowed them to move from soil shelters and hollow logs to a variety of more exposed nesting places, such as on leaves, on branches, and on rock faces. Since many paper wasps harvest and feed on caterpillars of various sorts, their ability to build nests directly on leafy plants allows them to live close to their food sources.

Ants are wingless social wasps that, unlike the bees and paper wasps, have more successfully abandoned the ancestral behavior of stocking single cells with food, and raise their young in larger com-

FIGURE 9.3. Face and tusk-like mandibles of a soldier caste of a New World army ant, *Eciton hamatum*, from Costa Rica. Ants such as these are formidable predators in Neotropical forests.

munal chambers. In this regard, the nests of many ants are more similar to those of termites. Ants also moved out of the soils to occupy a variety of plant habitats; many nest in hollow stems, inside nuts or large seeds, and even on open branches by tying together leaves with silk. Some ants, such as the infamous army ants of South America, have abandoned natural shelter nests entirely and, when needed, form temporary camps or bivouacs from the intertwined bodies of their own workers. The ants have diversified their diets more than any other group of social insects, and eat other bugs, plant materials, seeds, fungi, and sugary honeydew from a number of insects, including aphids, treehoppers, mealybugs, and caterpillars.

The social structure of the bees, paper wasps, and ants is very much like the one we considered earlier for their distant relatives, the termites: each of these lineages has evolved caste systems with workers, soldiers, and queens; each has overlapping generations of long-lived adults; and each cooperatively cares for their young. By convergent evolution, the termites and social Hymenoptera have also developed similar methods of chemical communication, complex nest structures, behaviors for exchanging liquid foods between nest mates, abilities to create and use chemical trails, and territorial behavior.[6]

On the other hand, termite and wasp societies have some major differences. Among termites, caste is determined by exposure to chemical pheromones, while among Hymenoptera, it is determined by nutrition and the type of food fed to the larva. Termites employ child labor, while bees, wasps, and ants typically do not. Because termites have gradual metamorphosis—their children are smaller models of the adults—the young are capable of working in the colony as they grow larger, and the adult workers take full advantage of the youth labor camps. The bees, ants, and wasps, by contrast, have complete metamorphosis: their young larvae differ enormously from the adults in form and possible functions. Social Hymenoptera larvae are essentially helpless grubs that exist as growing and feeding machines. For the most part, they can contribute nothing useful to society and do not begin working until the adult stage.[7] Also, termites have more equality of the sexes—there are both male and female workers—while the bees, ants, and wasps have female-dominated (Amazon) societies. There are many potential reasons for this, but an important consideration is that female wasps have a defensive stinger, while female termites do not. Female wasps also have a sperm-storage organ (the spermatheca) and the ability to fertilize—or not—each egg as it is laid. Since, as discussed earlier, in the Hymenoptera an individual's sex is determined by whether or not an egg is fertilized, bee, ant, and hornet queens mostly choose to produce female children. Presumably this is because female wasps are much more useful for hunting food or defending the colony nest with their stingers.

Termites evolved from a single common ancestor that became social, while social behavior among the Hymenoptera originated independently in multiple groups. Evolving the habit of building and provisioning nests certainly played an important role in the Hymenoptera's transition to sociality. A solitary wasp mother must leave her nest unattended while looking for food, thereby opening it and her offspring to predators and parasites. Even a very small social group benefits from allowing some individuals to guard the nest entrance while others search for food or building materials. Another important factor must have been the evolution of the Hymenoptera's defensive stinger. The stinger is a powerful tool not only for paralyzing and subduing prey but also for defending a growing nest against raids by larger animals, which in the Cretaceous years might have been dinosaurs, birds, or small insectivorous mammals. But perhaps the most fascinating aspect of social wasp evolution is the intriguing possibility that the wasps may have been genetically predisposed to becoming social because of their haplodiploid sex determining method.

For many years, one of the great questions about the social insects was why they evolved sterile workers. Charles Darwin pondered this puzzle of natural selection, but he was unable to solve it. If worker bees, ants, and wasps are sterile, they do not contribute any of their own offspring to the next generation. Why would any organism take care of someone else's children and not leave any of their own? A potential answer to this seeming dilemma, known as the kin selection hypothesis, lies in the curious genetic implications of the Hymenoptera's haplodiploid method of making males or females, whereby unfertilized eggs develop into haploid males, and fertilized eggs develop into diploid females. In the ants, bees, and wasps, sisters share more genes with each other (75 percent) than they share with their own daughters (50 percent). If, for example, a female worker ant were to develop her own ovaries, mate, and produce her own children, she would on average share 50 percent of her genetic makeup with each of her daughters. On the other hand, when female workers help their mother raise more of her daughters (more sister workers), they are helping to raise more individuals that share on average 75 percent of their genetic makeup, since these workers have the same mother and father. As strange as that sounds, it may help explain why the Hymenoptera have produced so many successful societies.[8]

The proliferation of social wasp societies during the Cretaceous years, but not earlier, is likely due to their tandem coevolution with the flowering plants. Bees totally depend on flowers for pollen and nectar, and in turn, they have promoted the flowering plants' success by impelling them to evolve and maintain successful pollination systems. The paper wasps might have gained their paper-producing ability somewhat earlier, but they needed to provision their nests with meaty food. Social evolution flowered figuratively while flowers bloomed literally, precisely because the angiosperms bred a cornucopia of edible insects, providing not only the wasps but also the ants with abundant provisions. The ants might well have evolved sooner, but their rise was advanced by the flowering plants' wide-ranging nesting opportunities—and by their nutritious seeds, which the ants, in turn, began moving and dispersing, thereby shaping Cretaceous plant communities. The angiosperms also stimulated a surge in homopterous insects—treehoppers, aphids, mealybugs, scales, and other relatives—whose sugary excretions supplemented the ants' diets.

Everybody Must Get Stung

The demise of *Tyrannosaurus rex*, *Triceratops*, and the other big dinosaurs is another of those lingering murder mysteries. While less catastrophic than the end-Permian mass destruction, the end-Cretaceous extinction has spawned even more solutions and public attention. The reason, perhaps, for this scrutiny lies in the charisma of the great dinosaurs and the implied importance of their extinction to our own species: it presumably opened the pathway toward the diversification of the mammals and the ultimate evolution of human consciousness. And yet all our big-brainy attentiveness to this event has not solved the mystery.

Among the many explanations for the dinosaurs' decline and fall is the idea that Cretaceous plant toxins poisoned them. However, it seems unlikely that these toxins would have been so pervasive as to drive the dinosaurs to extinction. Just as in the modern world, there would have been plenty of nontoxic plants and plenty with variable toxin concentrations. Herbivorous dinosaurs should have been capable of nibbling leaves and rejecting bitter or poisonous plants, just as herbivorous mammals are apt to do. It's certainly possible that as Cretaceous plant communities developed greater flowering plant diversity with more toxins, the dinosaur communities became more specialized and possibly less diverse and abundant. The angiosperms may have contributed to the decline of big dinosaur herbivores over a long period (millions of years, perhaps), but it is very unlikely that they were poisoned to extinction in a short time.

Some scientists have suggested that new kinds of insects, which might have spread deadly diseases, evolved. We have no clear evidence of this, but we can be fairly certain that the Cretaceous dinosaurs had to contend with small blood-sucking flies, and that the feathered dinosaurs, including some of the raptors, very likely had lice, probably parasites similar to the bird lice. Modern birds, reptiles, and mammals are attacked by various kinds of blood-feeding insects, which carry a number of different illnesses, so it is difficult to reason why dinosaurs would be immune to such assaults. It seems likely that they would have been bitten by these insects since at least the Jurassic years, and that epidemics would have soon followed, causing dinosaur population numbers to rise and fall. Maybe disastrous epidemics annihilated some species; however, there is no good reason to imagine that a single disease would wipe out all the dinosaurs at the end of the Cretaceous period, and yet somehow spare the birds and mammals.

The stinging social wasps, ants, and bees surely disturbed the dinosaurs during the Late Cretaceous. These increasingly vicious insects would have been crawling all over the forests. The vegetarian dinosaurs would have gathered their leafy salads ever more carefully, trying to avoid both plant toxins and, more serious, getting stung, which could have been lethal. The long-throated ones might have been more vulnerable, perhaps, but any browsing animal would be at serious risk of suffocating if they accidentally ate a wasp and were stung inside their throat. The stinging insects would have posed a choking hazard to even the gently nibbling duck-bills (hadrosaurs). And none of the stinging insects were happy about having their nests disturbed. Modern species are capable of holding their own against invaders, whether bird or mammal; because they first evolved alongside Cretaceous dinosaurs, some of which were likely to be at least partly insectivorous, it seems reasonable to suppose that stinging social wasps perfected their venoms in large part to defend themselves against nest-marauding smaller dinosaurs. Gone were the Triassic insect happy meals. The raptors, and perhaps even *Tyrannosaurus rex*, felt the wrath of the social wasps and the ants; they would have had to finish their meals quickly or risk getting stung by these opportunistic insects quick to scavenge bits of meat from animal carcasses. Has a hornet ever visited your picnic table?

It's possible that the venomous insect societies contributed to the dinosaurs' decline. But again, it's difficult to imagine more than that, unless the dinosaurs had wasp venom allergies, which is something we can never know. As entertaining as it may be to picture them gasp-

ing, swelling, and choking from wasp stings, I think it's far more likely
that the stinging insects simply made the dinosaurs' final years really
miserable.[9]

When Worlds Collide

It's not like we need any more possible explanations for the dinosaurs'
decline.[10] Many reasonable hypotheses exist, but they all boil down
to terrestrial and extraterrestrial causes. On the terrestrial side, the
main ideas involve either biological interactions (with plants, insects,
mammals, or other dinosaurs) or physical factors (climate change,
continental drift, or volcanic eruptions). On the extraterrestrial side,
which includes assorted ideas regarding meteors, supernovas, and
cosmic rays, unless you have been living in a cave with the wolves for
a few decades, you are probably aware of the widely popular asteroid impact hypothesis. This idea was championed by Luis Alvarez during the 1980s mainly because of the discovery that the rock boundary
layer between the Late Cretaceous and the Early Tertiary years, the so-
called K-T boundary, is loaded with unusually high concentrations of
the relatively rare element iridium. High levels of iridium have indeed
been found in the K-T boundary layer at many sites around the planet.
As far as we know, they could come from only two possible sources:
either the element was released from an impact with an extraterrestrial object, such as a large asteroid or a comet, or from deep in the
earth's crust by massive volcanic eruptions. The asteroid-impact hypothesis has gained additional support with the discovery of a potential impact site in the Gulf of Mexico, off the Yucatan coast. Although
some good geological evidence suggests that Late Cretaceous volcanic
events caused the dinosaurs to go extinct, when you get right down to
it, the asteroid-impact hypothesis is just plain sexy.[11] The idea that a
huge rock slammed into the earth, resulting in global mass extinctions
and general chaos . . . that's a real attention-getter. And while there
really is good evidence that the earth did get slammed by a big rock at
the end of the Cretaceous, let's not forget that we still can't clearly see
the actual extent of its damage. Truthfully, we will likely never know
for sure if the last *Tyrannosaurus* died from an asteroid impact or by
inhaling toxic volcanic fumes, or if it ate a poison flower, was stung

by a bee, caught the bird flu, or simply died peacefully of old age in a meadow full of flowers.[12]

The actual effects of an asteroid impact would vary according to the size of the object hitting the earth, but the potential effects could be devastating. Locally there would be an explosion, debris, heat, and toxic gasses. A really large impact would create a massive plume of airborne particles that could block sunlight and produce global climate change. If there is a lesson to be learned, perhaps it is this: as with the end-Permian extinctions, once again the effects of global disaster seem to be hardest on the big land creatures, as well as on marine organisms. Some insects (and plants) disappeared at the end of the Cretaceous, but unlike the end-Permian extinctions, no insect orders were lost. Some species decline occurred, but diversity increased again over the following millions of years. We need only look around our modern world to see that the insects, along with the flowering plants, the birds, and even the small shrewlike mammals (from which we fortuitously evolved), all managed to survive that asteroid impact. Even during the most catastrophic times, life is remarkably tenacious and resilient, especially if you are very small and have six legs.

The Cretaceous world was a beautiful place filled with flower meadows, butterflies, bees, duck-billed dinosaurs, and birds. One day that beauty was disturbed by a violent asteroid. We don't know for sure if the big dinosaurs died on that day, or lingered on for years or centuries, but we can be certain that somewhere in North America, roughly sixty-six million years ago, the very last *Tyrannosaurus* did die. Exhaling a final breath of air from its voluminous lungs, the King of the Thunder Lizards was no more. Some nearby beetles and flies likely fed on its heroic fallen body. For me, that death, more than any asteroid impact, best symbolizes the end of the Mesozoic era.

10 Cenozoic Reflections

1:05 AM. Sounds of Quito: the constant drone of auto engines, horns tooting, people talking, shouting, whistling, yelling, engines revving, sirens toot and wail, planes overhead, the constant rhythm of human street noise above the taxis, the accelerating buzz of a motorcycle, police whistles—a car alarm—pulsing . . . .

NOTES FROM MY FIELD JOURNAL

Quito Reverie
The pulsating stream,
Of paired crimson tail-lights,
On taxi cabs, flows, one-way,
Diagonally up the Avenida,
Like red blood cells, in an artery.

The conduit—thick with cars at midnight,
Nearly bumper on bumper,
Now flows loosely at half-past one.
A kilometer in the darkening distance,
The red tails slow and clump,
Brighten briefly, braking,
Then turn left into ebony oblivion.

Kilometers beyond,
The darkness deepens,
But on the rise of Andes mountains,
Can be seen the glimmering star-like lights,
Of human homes and sprawling city streets,
Dappling the hillsides with a twinkling spray,
Of mostly amber and white dots,
Interspersed with an occasional dull throbbing red,
And piercing bright blue.

These human lights are not stars,
And yet they resemble nothing more,
If not the sweeping pathway of a wide-splayed galaxy,
Mostly clumped in elongated array.

Yet, far beyond, in the utmost distance,
The furthest lights are sparsely grouped,
In the highest forested slopes,
Like angular, cryptic constellations:
Seven up—three across—two down.

Ironically, in the deep night sky above,
Not a single star or planet is visible.
The lights of space are totally obscured,
By dense Andean fog,
The haze of auto exhaust,
And a pale reddish glow of reflected city lights,
Like clouds of inter-stellar gas,
At the dawn of the Universe.

penned at the Hotel Rio Amazonas, Quito, Ecuador

Outside my hotel room lies Quito's urban sprawl, enveloped in the Andes Mountains' verdant, cloud-wrapped peaks. All of this starkly reminds me of the Cenozoic, our most recent geological era. Within the last sixty-five million years, tectonic forces uplifted the Andes, ultimately redirecting the water flows of South America and creating the Amazon River basin. The great diversity of modern neotropical plant and animal life was also shaped over the Cenozoic, and I'm in Ecuador with my students to search for previously unseen and undiscovered microscopic insects in the Yanayacu cloud forest—one of the most species-rich places on earth. Here in Quito, though, the urban chaos reminds me of another significant Cenozoic event: the evolution of the human species.

Were it not for the necessity of commenting on our origins and our impacts on this planet, I might not have bothered to write this chapter at all. You see, from my peculiar point of view as an entomologist, the mammals' recent history seems almost like a trivial aside to the insects' deep history, which, as we have seen, began hundreds of millions of years ago. Here, however, is my condensed version of the story of us. One day, when the dust from the Late Cretaceous, dinosaur-extinguishing asteroid settled and the earth recovered her biological rhythms, some fortunate shrewlike mammals—insectivores, mind you—scrambled in the comfortable leaf litter, hunting for grubs. For the first time in 150 million years or so, they were *not* being hunted by the dinosaurs. Let loose from the bondage of dinosaur teeth, these small, furry, nipple-bearing, and milk-producing rodents proliferated. Some of these, the early lemurlike primates, evolved to occupy trees. They still ate insects, but some presumably expanded their diets and consumed fruit along with other plant parts. So it went for tens of millions of years, as our ancestors crawled from branch to branch, nibbling on bugs and plants. Then, between 5 and 12 million years ago (during the Miocene epoch), the climate of eastern Africa changed. From fossil pollen records we know that the forests in that region became sparse and grasslands more widespread. We suspect that some of the arboreal primates climbed back down out of the trees and started foraging, more omnivorously, for food on the ground.

Because many of our cultures still eat insects, we can assume that our highly omnivorous hominid forebears did so as well.[1] They probably depended on bugs for protein just like chimpanzees, our nearest living genetic relatives. At certain times of the year chimps can spend up to seven hours a day feeding on termites, for which they fish by plucking a grass stem and inserting it into a nest entrance. The termites bite the stem then are easily extracted and eaten off of it (they're protein pellets on the cob). Fishing with grass stems is probably the simplest imaginable form of primate tool use, and I think we can also assume that our ancestors shared this curious behavior with the chimps.[2] In my opinion, the origins of human tool use, fine motor skills, manual dexterity, and ultimately the rise of human civilization are firmly rooted in our ancestral insectivorous diets. We may owe our very existence to social cockroaches. If termites weren't abundant, would primates ever have come back down out of the trees? I doubt it.

The Cenozoic years are popularly called the Age of Mammals, and you can probably guess why: because *we* are telling the story.[3] When humans discovered their own evolutionary history, the self-realization was almost too much to bear. The notion that we were not put here on a pedestal to rule the planet, but that we instead emerged from a long series of random and quirky events—that is a lot to contemplate.

What if Cambrian predatory arthropods had hunted our lowly ancestor, *Pikaia*, to extinction? What if Ordovician predatory sea scorpions, trilobites, and squids had eaten all of the early fishes? What if those scuttling Silurian and Devonian scorpions and centipedes had been so deadly as to prevent any vertebrates from colonizing land? What if during the Carboniferous the aquatic immature stages of the giant griffenflies had killed off all the immature stages of pond-dwelling amphibians? What if the Permian protomammals had been one of the casualties of the period's mass extinctions? What if, at any time during the hundred million years of Mesozoic dinosaur domination, those toothy predators had managed to catch and eat the last of the little mammals? What if the Cretaceous asteroid had missed earth and the dinosaurs continued their reign of terror?

In each case, the answer is the same: if events had gone slightly differently, humans might not be here at all. The fact that we are here to ponder these things is truly awesome, but it is certainly not inevitable. We not-too-subtly reinforce our fragile human egos by continuing to call the Cenozoic era the Age of Mammals. However, mammals are indeed just one unlikely sidebar on the history of life. The Cenozoic world was still very much dominated by the insects and flowering plants. By focusing on the mammals we distract from what I see as the true success story of the most recent era: the origins and complexity of tropical forest ecosystems and their domination by insects.

Fifty Shades of Green

Notes from my Yanayacu diary:

We live in the Age of Tropical Biodiversity. The tropical forests of this Cenozoic earth are replete with thousands of flowering plant species and millions of insect species. Now I have been in the midst of this majestic biodiversity for one week. The forest at Yanayacu is a remarkable contrast to the street scenes in Quito. Here, during secluded moments deep in the cloud forest, one can almost imagine what the world might have been like had humans never arrived. Every day it rains at Yanayacu. Especially in the afternoon, and late at night, the raindrops pound the tin roof of the research station, sounding much like Wyoming hail on the metal roof of my beat-up old minivan. Now it is mid-afternoon on a quiet and unusually clear day. From my vantage point in the station loft I can see, encircling the valleys, the moist forested hillsides of the Andes's eastern slopes. The sounds of water splashing over rocks and gushing through narrow passages, dripping everywhere and splashing on leaves—the sounds of water along the stream trail are indelibly imprinted in my auditory memory. Water is everywhere and ever-present at Yanayacu. So much so, that the station dog is named Rain. Even in the rain, at night the blacklight sheet draws in thousands of moths. They cover the windows and walkways in the morning. The hyperdiversity of moth species at Yanayacu rivals that of any place on this planet.

A hike down the Yanayacu Stream Trail is a great way to appreciate Cenozoic tropical diversity. The trail snakes along the stream, crossing it many times. Sometimes we ford the stream on slick mossy stones, or wade in the cold mountain water. Sometimes we cross like acrobats on slippery algae-covered trees or boards laid loosely across the chilly water gap. On particularly rainy days, the stream rises, and you are sure to get plenty of cold water in your boots. Even on the trail the path is slick with mud and the way is sometimes treacherous. Above the water noise, sometimes we hear the piping and trills of unseen tropical birds—the most successful feathered Cenozoic dinosaurs—hidden in the dense foliage, or high in the forest canopy.

The forest is as green as it is wet. Even in the dim light of the understory, the forest is painted with riotous shades of green. Bluish-green to yellowish-green, lime green to olive, grass green to sea green, green in fifty shades and more, the color of an endless tropical summer, verdant and lush. Verde, green, the color of plants, the color of life: an ever-present reminder of the pigment chlorophyll, the stuff that allows plants to capture sunlight's energy. Green, a constant reminder that our sun, our nearest star, provides this radiant energy for our terrestrial ecosystems.

Competition for sunshine fundamentally shapes the forest. The trees grow tall and spread their upper limbs to make a dense canopy, competing to filter much of that light in the highest layers. The light level at the floor is quite low, even at midday. In a mature tropical forest, 90–95% of the sunlight is filtered away by the plants before it can reach the ground. Some understory plants grow huge leaves, many feet across, to better soak up the filtered sunbeams. In small gaps in the forests, where old trees have fallen, more intense light briefly reaches the surface. Young trees sprout rapidly from huge seeds that provide concentrated nutrition for the race upward. Most are wrapped, entwined in clinging vines seeking to make the same journey. The competition for sunlight, space, and nutrients is fierce, and only a few young seedlings will survive and mature to old forest canopy trees. Small epiphytic plants, like orchids and bromeliads, cannot hope to compete with the towering trees, so they stand on the giants' shoulders: by sitting on branches, they get closer to the precious sunlight above. The density and weight of these epiphytic plants eventually becomes so great, it is not unusual for huge moss and orchid-laden branches to crack, and come crashing to the forest floor.

Along the Stream Trail the tree trunks soar high, like the pillars of a green basilica—the basilica of life. Quito is replete with human-built cathedrals, hundreds of years old, but I prefer the natural forest cathedral, millions of years old. All the trees are heavily matted with bromeliads, algae, lichens, and mosses. At close range, the surface of almost any log looks much like the mossy shoreline of a Silurian age stream—that miniature plant community where insects had their earliest terrestrial origins. These mosses still crawl with millipedes, centipedes, and microscopic wingless insects, as they have since the Late Silurian and Early Devonian. Along the path in the forest understory are ten-foot splayed tree-ferns, ancient survivors, reminiscent of the Devonian age forests where true insects first crawled, and later sprouted the earliest wings.

At this high elevation, it is easy to imagine and appreciate the solar-panel hypothesis for the origin of insect wings. It is cold here, even though we are situated nearly on the equator. The elevation, the forest shade, and the ever-present water conspire to put a definite chill in the air by day. At night it is even more surprisingly cold. Passing the tall tree-ferns, I visualize Devonian-age insects basking themselves on the upper fronds. We have seen some of them today: machilid jumping bristletails on logs and stones, and primitive wingless diplurans under rotting logs and leaf litter—only slightly modified survivors of the ancient times.

Pausing at a stream crossing, I see taller trees arching over the rivulet; a storm has dislodged one into the stream, from its eroded banks. The water is clotted with decaying, algae-matted, woody debris. I linger here to meditate on the Carboniferous times, and to visualize the masses of plant materials inundating the ancient waterways. At Yanayacu you can touch, smell, and feel the rotting plants, and maybe get a bootful of dark, tannin-tinted mud, suitable for fossilization—if only the insects, worms, algae, and bacteria don't get to it first. It is easier to imagine the Carboniferous forests at this elevation, because here we are too high up for many termites, and the fallen mossy wood accumulates deeply. Lacking termites, as in Carboniferous days, the storm-wracked trees at Yanayacu, always dripping with water, are more slowly eroded by bacteria, algae, fungi, lichens, and even gigantic tropical earthworms. My coal-swamp reverie is gently broken by the flight of a damselfly along the streamway—an apt reminder of the Carboniferous insects that evolved large wings and could fly. The fair damsel pauses briefly on a fern frond until, disturbed perhaps by a raindrop, she continues her flight upstream. I break from my Carboniferous reflections and continue trekking downslope.

As we make our way along the trail, I notice further that Yanayacu cloud forest is a mosaic of past survivors. We see the modern relatives of the Permian neuropterans, lacewings, and beetles. We see the descendants of Mesozoic sawflies, and the great-great-grandnephews of feathered dinosaurs still flit and twitter in the treetops. But the Cretaceous innovations—the social wasps, ants, flowers, and especially the exposed-feeding caterpillars—greet us riotously.

Slipping briefly in the mud, instinctively catching my balance like a soggy Tai Chi master, I flash from my Mesozoic daydreams back to the present, and continue my Cenozoic reverie. We follow the watercourse now along slopes that were uplifted during the Cenozoic years. The chilled waters gurgle down toward the Amazon basin—but that is not our destination. For on the Andes's eastern slopes, the insects will come to us. Over the course of millions of years, pulses of Amazonian insect species have migrated to the highlands and adapted to its cooler climate.

My destination is the forest at the base of the stream trail. I reach the site with considerable anticipation and excitement. Anticipation, because near here, two years earlier, my graduate student, Andrew Townsend, fortuitously sampled a specimen of a new euphorine wasp species, which belongs to the group of microscopic parasitic wasps that has become the focus of my life's research. Drew did not realize his discovery at the time, but I later found the new wasp among his samples when we studied them under microscopes in my remote Wyoming laboratory. My excitement doubled because the specimen was not only a new species, perhaps never seen before by human eyes, it was also a new genus.

I have returned with my research team to search for more undiscovered insects. I am hoping to find more specimens of my new wasp genus, so we can better understand its variation, but I know that the task is daunting and has little chance of succeeding. Finding one rare three millimeter long wasp in the Ecuador cloud forest is far, far more difficult than attempting to find a needle in a haystack. If only it were that easy.

We come armed with the knowledge of precisely where and when the wasp was collected, and we know a simple sampling method that might work: yellow pan trapping—the very method Drew used to collect his wasp. Bright yellow plastic bowls are placed along the forest trails. They are filled with slightly soapy water (in this case I'm adding a surfactant, the same stuff you put in your dishwasher to remove water spots), to reduce the surface tension so that tiny insects will more easily fall into the fluid. Flying insects see the Day-Glo yellow bowls as bright spots on the forest floor, and because many of them are strongly attracted to the color, they dive right in. In a dry climate like Wyoming the bowls would dry out quickly and need to be replenished. But at Yanayacu cloud forest the ever-present raindrops keep them filled. You just need to scan the bowls frequently to pick out promising specimens, or filter them out with a fine-mesh net.[4] Drew placed 20 bowls and found one golden nugget. I'm here to up the ante by placing 200 along the same trails. But will the method work again? I am troubled by the thought that very likely the insect I'm searching for lives most of its life high up in the forest canopy. I may be hiking directly under thousands of them. Perhaps two years ago that one particular wasp randomly spotted a yellow dot on the floor and flew down to investigate. Perhaps decades might pass before anyone sees it again.

Rise of the Imagobionts

My worries, while realistic, proved to be unfounded. After four days of hard work we managed to find one more specimen of the new species, and by the end of the week I was very pleased to have found three of

FIGURE 10.1. A solitary male of a parasitoid wasp, *Napo townsendi*, perches on a *Dendrophorbium* leaf in the Yanayacu cloud forest in Ecuador. (Photo by Andy Kulikowski.)

them. The field station manager, Jose Simbaña, had also been running yellow pan traps for ten days a month, every other month, for the last year. His trap samples consisted of pint-sized jars full of tens of thousands of microscopic insects, pickled in alcohol. By day we checked my yellow pans, and by night we worked in the laboratory, sifting through Jose's samples one spoonful at a time. We looked at so many samples under the microscope that late at night I could close my eyes and see the burned-in images of insects floating in a sample dish. At the beginning of the week, I joked that when I found my wasp, everyone would know it because I would shout "Eureka!" By the time I finally found one, I was so physically and mentally exhausted that all I could manage was a low whistle and "Here it is." I'd estimated that we looked at more than a hundred thousand other insects for each one of my new wasp species that we spotted. Why are some insects so numerous, while others are extremely scarce?

Different insect species have different behaviors that affect their population levels and relative abundance. In a forest like Yanayacu, good examples of very common insects are the fungus gnats. The maggots of these tiny flies breed in enormous numbers in the forest floor's decaying leaves. The flies are everywhere, and they dive into the yellow pans like crazy. If you wanted to, you could easily sample hundreds of them in a morning's work. Other insects, such as ants, are social, so again it's not difficult to sample large numbers of them.

My new wasps, on the other hand, are solitary insects that belong to a larger group (subfamily) of microscopic wasps known as the Euphorinae.[5] They are also a special kind of parasitoid whose unusual biology contributes to their scarcity: they live and feed as immatures inside the bodies of other adult insects, such as beetles, bugs, bark lice, green lacewings, ants, bees, and even other parasitic wasps—a difficult kind of parasitism that I call imagobiosis.[6] Adult insects are the least abundant life stage for any insect population, and are therefore harder to locate, compared with eggs or larvae. An adult insect is densely armored and may defend itself by biting, kicking, or spitting chemicals. It is fully mobile, so if all else fails it can run or fly away. How did imagobionts evolve to overcome these obstacles?

In our previous discussions of parasitism, we considered idiobiosis, where venom is injected into the host insect, rendering it permanently paralyzed. Idiobionts tend to attack hosts that are embedded in plant tissues or wood (and are therefore barely moving in the first place) and alter them so they can't move any more. That strategy for living and feeding worked really well when it was first developed back in the Jurassic days, and it still works well for tens of thousands of modern parasitic wasp species. But the idiobionts are pathetically slow at laying their eggs. A wasp that attacks beetle larvae concealed deep in wood commonly spends the better part of an hour or more just laying one egg. It needs to drill a hole through dense wood, insert its ovipositor, locate the host beetle, inject paralyzing venom, lay an egg, then withdraw its ovipositor from the plant substrate. Idiobiont wasps apparently could not evolve to attack mobile adult insects directly. An adult would get fed up and fly away before the wasp could get the job done. The evolution of the ability to parasitize adult hosts required koinobiosis.

Koinobiosis, you may recall, is a style of parasitism where more exposed, abundant, and mobile hosts are injected with eggs, and where larval development is delayed until the host becomes much larger. The more modern and common style of parasitism practiced by thousands of wasp species in this planet's forests, it allows koinobionts to successfully find and eat a much broader array of insect species, especially young caterpillars feeding on plants. Since they tend to attack insects that are exposed or only shallowly concealed in plant tissues, koinobiont wasps evolved very fast egg-laying abilities. One of these wasps can run up to a caterpillar, stab it with its saberlike ovipositor, and inject an egg into the caterpillar's body within several seconds. This rapid attack set the stage for adult parasitism.

The imagobiont euphorine wasps have evolved even faster egg-laying abilities. They fly or run at their adult host; slip their razor-sharp ovipositor through its membrane, between its exoskeletal plates or into its anus; and blast an egg into it, all in just a fraction of a second. An imagobiont larva develops immediately, often killing the host, and emerges in a few weeks.

The Million Bug March on the Forest

Why would an insect bother to live inside another adult insect, when there are so many more less-mobile and easier-to-attack species of caterpillars and other soft things to eat? After all, by fifty million years ago an exceptional variety of immature insects were feeding on plants, probably more species than had lived at any previous time; however, the idiobiont and koinobiont wasps were equally successful, so much so, that they would have constantly competed for those plant-feeding hosts and hammered them with parasites. We see that struggle today, in a forest like Yanayacu, where any given caterpillar might yield as many as ten or fifteen different parasitic species. But when imagobionts came along and started attacking adult insects, no other species was competing for the available food inside them. As a result of invading this entirely novel adaptive zone, the adult-parasitoids diversified rapidly.

When organisms adapt to colonize previously unoccupied niches, these unique opportunities sometimes allow large numbers of new species to evolve—a phenomenon called adaptive radiation. The Cenozoic earth provides several nice examples: the radiation of mammals into niches formerly occupied by the dinosaurs, of flowering plants and pollinating insects, of plant-feeding caterpillars on new species of flowering plants, of parasitoid wasps and flies on caterpillars and other insects, and of the epiphytic plants, especially orchids and bromeliads, into the structural complexity of forest canopies. The birds, mammals, and amphibians radiated in response to insect diversity and the availability of new niches: for example, the evolution of water-collecting concave leaf bases in bromeliads provided new habitats for frogs and aquatic insects. The rise of the insects, along with fruiting plants, promoted diversity in a particular new mammal group, the bats. Evolution of bats and other mammals, and the birds as well, provided niches for the radiation of many new parasitic insects, particularly species of bird and mammal lice, as well as fleas. Bats don't have lice, but the niche was occupied anyway by another new group: parasitic bat flies. The history of the Cenozoic teaches a simple lesson about the nature of tropical ecology, and the nature of life in general: diversity promotes diversity. When species live together and interact, multifarious new behaviors evolve. As time goes on, more species evolve, and the more complex and interesting biological diversity seems to become.

Adaptive radiation isn't unique to the Cenozoic Period. Looking back over the history of life, we've seen some great examples of how natural selection drives it, and how new organisms diversify to occupy vacant ecological niches. Precambrian times showed us the expansion of microbial life into nutrient-rich oceans. In the oxygen-rich Cambrian we saw the radiation of multicellular respiring life, which filled the oceans with various exoskeletal animals. The Silurian showed us the invasion of land in response to ozone enrichment and the filtering of harsh solar radiation. Silurian plants and animals enjoyed an adaptive migration into shoreline niches, creating the earth's first terrestrial communities. Devonian times gave us the expansive evolution of land plants inland and upland away from shorelines, and diversification of the insects with plant communities. The Carboniferous showed us the rapid ascent of insects with wings and the invasion of the air. Permian times were marked by the explosive evolution of insects with complex metamorphosis into numerous, previously unoccupied niches. It also showed us the single biggest setback in the history of life. But life, especially insect life, proved itself resilient, at least over the long term.

It's tempting to look back over the history of insects and view their diversity as steadily rising. If we were to plot this over time, starting from the origin of the first true insect about four hundred million years ago to the present day, our trend line would rise from one to seven million species, to pick a modern but conservative estimate—the number might well be higher, perhaps as high as thirty to fifty million, according to Terry Erwin. That would be a fair approximation of insect species accumulation, but the actual pattern wouldn't be perfectly linear, because not all species evolve into new ones at the same rate, and as we have seen, sometimes extinction occurs. The end-Permian event impacted insect diversity more than the end-Cretaceous event did, so the decline in species at that time would have likely been the biggest dip of all. We can't be sure about the exact numbers of species during each geological period, just as we are not sure about the current number, since the fossil record is too incomplete. But that trend line from one to seven million gives a good coarse view of what insects accomplished over four hundred million years.

Tragedy of the Un-Commons

It is tempting also to imagine that we live in the very time when insect variety, as well as that of other life on earth, has achieved its highest levels. Yet we must consider that the actual peak of living diversity was reached earlier in the Cenozoic, maybe two hundred thousand years ago, before humans extensively changed the planet. To assess our impacts on insects, we need to briefly reconsider the story of us. By twenty thousand years ago, we were probably starting to affect species diversity through fire use and hunting. With the onset of agriculture, around ten thousand years ago, we altered the natural landscape even more. This was soon followed by cities and warfare, and the resulting depletion of natural resources no doubt took some toll on small plants and animals, especially insects with narrow distributions. Over the last four hundred years or so, the advent of the industrial revolution and medicinal advances accelerated human population growth, further increasing our influence on the natural world. Over the last hundred years we have gradually become aware that humans are causing extinctions, or the threat of extinctions, of many animals and plants, not just by traditional hunting practices but merely by expanding our own habitats into theirs.

We are currently living in the midst of another extinction crisis, one especially due to the rapid destruction of tropical forests, where the largest proportion of biodiversity is found. Since the late 1980s this has been dubbed the "biodiversity crisis," a phrase popularized by Edward O. Wilson. We now realize that much of the earth's biological resources are at severe risk. Insect species, especially uncommon and rare ones living in the tropics, are particularly at risk of extinction, since many have narrow distributions and occupy small, highly vulnerable niches. One published study in the 1990s estimated that the rate of extinction for microscopic insects may be as high as one to two species lost per hour. For the most part, these are very small species, many of which have never been studied or named.

The estimated loss of species is such that over our lifetimes, a large portion of this planet's biological diversity may be lost forever. Sure, there have been many extinction crises in the deep past, as well as many smaller episodes. But for the first time in nearly four billion years of life, a massive extinction event is being caused by the global spread of one single species, *Homo sapiens*. The tropical forests may be the most severely affected, but our influence is so pervasive that we alter virtually all habitats on our planet.

One might try to rationalize this situation by thinking that there are millions of insects, so why would it matter if any of them are lost? Some might think, who cares if one fly or bee goes extinct, if that fly or bee belongs to a species that has not been cataloged? But what if that fly—or bee, or wasp, or beetle—were a specialized pollinator of an epiphytic plant? The fall of that one insect species might cause a subsequent loss of plant species, and also of other insects that depend on them. This pattern of the removal of one keystone species stimulating the loss of others is called a cascade effect, a real phenomenon that has been documented in the ecological literature.

We should remember that each living species is unique. Each fills a distinctive ecological role, and each encodes, in its genes, some qualities not found in other species. Some of those characteristics could potentially benefit future humans economically. Insects can be sources of oils, fibers, waxes, scents, even food. Possibly millions of species produce silk, but only a few have been domesticated and used commercially. Undiscovered insects might produce medicines that could cure human diseases. We cannot easily fly to another planet to find new, useful species, if they exist, so instead of asking whether or not we should care about the loss of insects, we should ask a more relevant question: how can we justify the current mass extinctions when we are wiping out potentially beneficial resources that can never be replaced in our lifetimes? If we could stop destroying habitats, many insects could survive, and many new ones would evolve in the distant future. Yet it would be unwise to simply wait around for all this to happen; we should instead conserve as many of the existing species as possible, in the hope that we'll eventually discover their potential benefits.

Setting aside the utilitarian perspective, living species are also intrinsically interesting simply because they exist. Yesterday I took a short hike in the Yanayacu cloud forest with six students, and in less than ten minutes we found a new species of the *Ilatha* wasp that I have been studying, one that attacks toxic *Altinote* butterfly caterpillars on a yellow aster plant called *Munnozia*. In the space of an hour we found eighteen specimens of the new wasp, and we spent fifteen minutes watching and photographing a female as she probed and laid eggs into a caterpillar. In the end, whether or not this particular species ever benefits humans really doesn't matter. It is unexpectedly fascinating, and, moreover, discovering a new species can be a powerful teaching tool for recruiting future generations of biologists. Now more than ever it is essential to educate young students about the diversity of life, especially since living species are going extinct faster than biologists can name new ones.[7]

It seems very ironic that, on the one hand, so many people are enthusiastic about the prospects of exploring space and other planets for new life forms, but, on the other, are blissfully unaware of the unexplored life here on earth. Where is the joy in sending a space probe to Mars and scraping the dust for remnants of ancient but now extinct microbes, even if they did once exist? It would be great to learn that microbes once lived there, but this discovery could wait a few decades. Wouldn't it be just as exciting to send robotic probes or, even better, well-equipped human missions to the forests of the Amazon basin to discover strange and unknown life forms? We would be able to examine seemingly limitless new species that might go extinct in fifty years if we don't act now to discover them. The public would be amazed at the "alien" creatures residing on our own little world, still the only place in the entire universe where we know for sure that life exists.

Astronomers, astrophysicists, and cosmologists find the universe's trillions of galaxies to be wonderfully complex and interesting, and they see the galaxies as the glorious outcome of the big bang. Crack open a cosmology textbook and you will probably find that the final chapters are about galactic diversity, and rarely about living creatures. As a biologist, however, I am compelled to see the living universe as the big bang's most awesome outcome. The anatomy and physiology of a single insect is in many ways more complex than the structure and function of a star. Likewise, the assortment of living things in a tropical forest, and the intricacy and complexity of their ecological webs, is more fascinating and elegant than the physical structure of any galaxy.

Astronomers and the public would be shocked if, one by one, galaxies began to wink out and disappear forever, especially if they were to vanish so rapidly that half would be gone in the next hundred years. What if we discovered that humans were responsible for this? Would the world stand by and allow us to extinguish galaxies, if we knew a way to stop destroying them? Everyone should be alarmed because we are daily losing our biological treasures, the living legacy of four billion years of organic evolution. Conserving a large portion of our living species isn't that complicated. All we need to do is stop eliminating habitats and start studying this planet's biology with a renewed vigor. If, during the next fifty years, we could promote biological education and boost research funding, we could save millions of species. The good news is that we don't need to build a spaceship and fly to distant stars just to discover new forms of life; we only need to book a ticket to Ecuador, Brazil, Peru, Indonesia, or dozens of other tropical countries and take a walk in the forest.[8] Or, if we continue destroying habitats and accelerating global climate change, we could expunge most of the remaining tropical forests along with myriads of species. The choice is ours.

As I finish this chapter, I am looking out over the Yanayacu Valley with a clear view above the dappled forest canopy, to the furthest ridges of the eastern Andes. It is not raining at the moment, the sky is brilliant azure, ethereal mists are rising from the treetops, and billows of white clouds are balanced on the farthest, highest peak, like massive cotton balls. As far as the eye can see, in any direction, lies the forest canopy. Viewed from above, it looks like endless fields of discolored broccoli clumps. The forest in the farthest distance appears more uniformly green, but in the foreground I can see the patchiness of the trees, each crown of which is distinct, with its own shape, colors, and community of epiphytic plants. The leaves of many tree crowns are greenish but also visible are shades of yellow, white, brown, pink, orange, and red. I am looking at hundreds of different tree species forming an intricate botanical mosaic. On the nearest trees, outside my windows, I can closely appreciate the voluminous epiphyte load of diverse bromeliads, mosses, and lichens. This valley alone must contain more living species than inhabit all of the eastern United States. Just beyond my sight, over the farthest peaks, lies the species-rich Amazon basin of Peru and Brazil.

As the sun sets, the swallows fly back to roost in the rafters above my room, and the stars begin to emerge in the quiet and misty night sky. As darkness falls, hundreds of moths begin to fly, and bats replace the insectivorous swallows. A storm rolls in over the distant Amazon, putting on a psychedelic heat lightning show over the furthest ridge. But above the Yanayacu Valley the sky is still clear. The Milky Way reveals itself with startling clarity and brightness. Some of the students sit quietly in the highest rafters, admiring the stars and meteors. Looking toward the stars, I can't help but wonder: does life exist elsewhere?

Postscript: The Buggy Universe Hypothesis

No reasonable mind can assume that heavenly bodies which may be far more magnificent than ours would not bear upon them creatures similar or even superior to those upon our human Earth.

GIORDANO BRUNO, *De l'Infinito Universo e Mondi*

While in Nantucket he had chanced to see certain little canoes of dark wood, like the rich war-wood of his native isle; and upon inquiry, he had learned that all whalemen who died in Nantucket, were laid in those same dark canoes, and that the fancy of being so laid had much pleased him; for it was not unlike the custom of his own race, who, after embalming a dead warrior, stretched him out in his canoe, and so left him to be floated away to the starry archipelagoes; for not only do they believe that the stars are isles, but that far beyond all visible horizons, their own mild, uncontinented seas, interflow with the blue heavens; and so form the white breakers of the milky way.

HERMAN MELVILLE, *Moby Dick*

We have every reason to believe that there are many water-rich worlds something like our own, each provided with a generous complement of complex organic molecules. Those planets that circle sun-like stars could offer environments in which life would have billions of years to arise and evolve. Should not there be an immense number and diversity of inhabited worlds in the Milky Way?

CARL SAGAN, *The Search for Extraterrestrial Life*

I'm beginning to wonder if we haven't been making an erroneous assumption about the history of life. When we learned that humans were the highly unlikely result of a long series of contingent and fortuitous events, perhaps we jumped to the false conclusion that evolution is never progressive and no kinds of animals are predictable. But just because one thing is unlikely doesn't mean that all things are equally unlikely. Instead, some events are more likely or less likely than others, as any insurance salesman can tell you. That's why we are required to purchase automobile and homeowner's, but not volcano, asteroid, or terrorism insurance. There may be only one human species, but there are tens of millions of insect species. This fact alone suggests to me the likelihood that insects are more probable creatures than anything else.

I also find myself wondering about the existence and nature of life on other planets, just as Giordano Bruno did centuries ago. And increasingly I find myself contemplating a particular heretical thought: can we expect some aspects of life's history on earth to be repeated elsewhere? I'm not claiming that events on other planets would follow exactly the same sequence, or that there are lots of other worlds with humanlike beings. Many modern scientists, most notably Carl Sagan, have asserted that other planets exist with their own races of intelligent, big-brained creatures, and many people consider this assumption to be so plausible that astronomers scan the skies for possible radio signals from other stars, and millions of Americans believe in aliens and flying saucers. It seems to me, however, not only that intelligent, big-brained creatures must be rare in the universe, just as they are on this planet, but also that small arthropod- or insect-like creatures are far more likely to exist elsewhere, since tens of millions of insect species live here on earth. For every single planet that should reach the "lucky" stage of evolution and spawn an intelligent race, surely there must be hundreds or possibly billions of habitable planets that never reach this stage but simply stabilize at a more probable phase of planetary evolution: worlds filled with oceans, oxygen-rich atmospheres, forests, and plants, which can hardly survive and diversify without insects or something very similar.

Let's consider what might have been happening simultaneously on other planets around other stars like our own sun while life was evolving on earth. Astronomers have determined that the critical components of living planets—carbon, oxygen, water vapor, iron, amino acids—are visibly abundant across the universe. Since there are billions of stars just like our own within the Milky Way and also trillions of other galaxies, each with billions of stars, many of which must be similar to our sun, we must assume that similar aggregations of matter would condense into habitable planetary systems in many places where other yellow stars exist.

Moreover, based on the observation that life on earth first evolved very quickly after our atmosphere developed, the planet cooled sufficiently, the liquid-water oceans formed, and the late bombardment of meteors declined, it is plausible, even probable, that life would evolve elsewhere under similar conditions and in terms of the simplest bacterial forms emerging. This idea has been widely embraced and pervades our scientific culture in several ways. It forms the rational basis for the science of astrobiology. We employ and provide grants to scientists who are engaged in the search for extraterrestrial life, even though no such life has ever been discovered, because we *do* consider it to be plausible, even likely. It's the reason why we are currently searching for evidence of fossil microbes on Mars and why we probed for evidence of living processes on Titan, the moon of Saturn with atmospheric conditions similar to ancient earth. The plausibility of bacterial life elsewhere is also the reason why most modern introductory astronomy books include a chapter on the chemical origins of life.

Although astronomy books will tell you how likely it seems for single-celled life to get started, they will tell you virtually nothing about the directions life may take beyond this point. For that perspective you need to review biology textbooks—but they will discuss the history of life on earth and often don't mention possibilities elsewhere. Nevertheless, by looking at this history, we may be able to surmise much about potential extraterrestrial patterns of life. One obvious conclusion is that single-celled organisms probably do not develop rapidly into multicellular organisms, anywhere. The biggest reason for assuming that life does not progress swiftly to complexity is the observation that bacterial life on earth remained single-cellular for billions of years, and it took more than three billion years for multicellular life and animals with rapid metabolisms to emerge. This slow pace of events certainly doesn't make the progression to complexity look inevitable, yet perhaps we can liken it to the slow pace of events in stellar evolution. A star like our own sun may appear to be constant or unchanging for billions of years, but eventually it will rapidly transform into a red giant because of conditions that were predictable at the outset. So assuming similar, earthlike chemical conditions on similar planets around stars of similar intensity, we have to assume that extraterrestrial life forms would evolve the capacity to split water molecules with starlight (the most abundant and predictable source of energy across the universe), form organic molecules with the released energy, and produce free oxygen. Photosynthetic life would, by its very nature, transform the atmospheric conditions on any planet. Oxygen reacts with iron in the oceans and rocks until stable compounds are formed, and once production of this gas exceeds the amount used in reactions with iron, any atmosphere would change and become more oxygen-rich. If the evolution of respiring multicellular animals is indeed a natural adaptation to avoiding oxygen toxicity, then an explosion of multicellular forms is an inevitable outcome of physically predictable processes.

It is one thing to suggest that oxygen might stimulate the evolution of multicellular complexity and a diversity of animal forms. It's another thing entirely to suggest that it might stimulate a progression to a particular form of animal: namely arthropods, but that's precisely what I am arguing. Aerobic respiration leads to more rapid metabolism, which promotes the evolution of fast-moving predators and more complex ecological food webs. Rapid metabolisms require that locomotion systems with articulating, muscle-manipulated, skeletal parts evolve. Skeletal systems can either be on the exterior or in the interior, and we have seen that external skeletons are easy to produce metabolically from excretion byproducts. It wasn't just luck that the Cambrian explosion produced a predominance of external skeletal forms—they are more effective for defense and therefore better for survival—and if we can learn any lesson from earth insects it is simply this: small is good. Small creatures require fewer resources to survive, they can occupy tinier and increasingly specialized niches, and they evolve more quickly than large organisms because of short generation times. So I'm arguing that the arthropods were not only a predictable product of the oxygen-generated Cambrian explosion but also the predictable colonizers and survivors over subsequent periods. Arthropods colonized earth's soils, forests, and air because they were the most diversified and best-adapted organisms to make those transitions. The very simple traits of small size, wings, and metamorphic growth ensured that insects were better colonists and survivors than any other organisms, and they proved themselves resilient to even the most catastrophic events, whether it was continental collisions, global climate change, massive volcanism, or asteroid impacts. For all of these reasons, I believe that small arthropod- and insectlike creatures would also be likely to evolve and survive on other planets, wherever the right conditions for life exist.

Perhaps you are skeptical about my idea. That's fine; skepticism is healthy for scientific thought. Some people might think that it isn't scientific at all—just a bunch of armchair speculation. Alright then, let's express my idea as a scientific hypothesis. I'll call it the buggy universe hypothesis, which, simply stated, goes like this: the living universe is full of bugs. I'm not predicting that other worlds are full of insects just as we have them on earth: butterflies, ants, beetles, flies, and such. Life elsewhere might have any number of possible variations. It might be based on DNA that spirals differently than our own or on other self-replicating molecules entirely, and it might have evolved multitudinous body forms, but on any planet where life develops beyond bacteria into multicellular animals, and where plantlike photosynthetic organisms colonize the land, the most abundant and diversified forms of life should be small exoskeletal creatures approximating arthropods. They might have different kinds of body regions, types of segmentation, and numbers of legs, but they will resemble arthropods nevertheless.

The buggy universe hypothesis is verifiable and has already passed one test: this planet is observed to be astronomically full of bugs. We can easily imagine other pathways by which life on earth might have evolved without any humans, or even without any mammals or dinosaurs, but given the unfolding of the earth's history as we understand it, it's difficult to imagine how terrestrial ecosystems could have evolved without insects or insectlike creatures. I can also conceive of three possible future tests. We might actually make contact with intelligent extraterrestrial beings. Once we learn to communicate with them, assuming they are friendly and conversational, we could ask them about conditions on their planets and so learn about extraterrestrial biology. We might also develop the optical technology to view conditions on distant planets. At present we can't even see those planets, so it's a stretch to believe that someday we might be able to resolve distant surfaces well enough to see plants and insects. Then again, when I was a kid nobody imagined that we would have satellites with optics capable of resolving license plates from outer space. The most likely possibility, in my opinion, is that we might discover and travel to other habitable planets. I'm skeptical about doing it the Star Trek or Star

Wars way. That is to say, I'm not so sure we will ever develop warp drives or hyperdrives or a way to exceed or even approach the speed of light. But with existing technologies we could devise ways of getting to other stars at slower speeds. Interstellar travel might take a long time, but humans could possibly traverse such distances using cryogenic hibernation systems or simply send robotic probes capable of conducting studies and returning data.

There's an old saying that goes, "You only hit a target if you aim for it." We have already proven that we can detect life-sustaining planets from a remote distance using the simple methods of spectral analysis and infrared photography, and it seems to me that these methods will continually enable us to search for planets with the traces of life known to us.[1] If astronomers (and botanists) find it plausible that plants might exist elsewhere—enough that we build spaceships with remote sensors to detect the presence of chlorophyll—then why shouldn't we consider it plausible that other planets might be highly diversified with arthropod- and insectlike creatures? The presence of chlorophyll across whole continents suggests not only the mere presence but a broad diversity of plants, and, as a biologist, I find it impossible to imagine how such wide-ranging plant systems could have evolved without the ecological interactions of insects, or similar arthropods. After all, arthropods affect soil quality and nutrient cycling, and, as we have seen, they also promote plant diversity through pollination, seed dispersal, and herbivory. The colonization of a planet by plants is not a solitary process—it is a complex symbiosis involving coevolution with soil microbes, fungi, and multitudes of small animals, especially insects.

My buggy universe hypothesis is contrary to much of what I was taught, but lately, I've been finding it compelling. At night, when I stargaze, I start to see insects and wonder if they do in fact live on other planets. When I look at constellations, I join those ancient people who saw arthropods and imagined Scorpio, the scorpion, and Cancer, the crab. The Big Dipper looks more like a long-tailed wasp, Draco the Dragon more like a millipede. Taurus the bull resembles a long-horned beetle, and the Gemini twins might be the eyespots of a saturniid moth. During these moments, I sometimes reflect on the biologist J. B. S. Haldane, who, when asked "what one could conclude as to the nature of the Creator from a study of his creation" reportedly responded "An inordinate fondness for beetles."[2] For decades, evolutionary biologists have treated his remark as a humorous anecdote, but I think we should take it seriously and contemplate why there are so many insects. And, if you are among the large number of people who believe in one creation story or another, you still need to reconcile your thinking with Haldane's compelling observation that this planet is full of insects. If God crowded the earth with bugs, then he must have done so for a reason. I'd have to conclude that, in his infinite wisdom, he would have made other planets buggy as well.

Maybe we never will discover life anywhere else. Or maybe, just maybe, someday we will detect another planetary system that shows all the signs of potential life. Perhaps we will devise a spaceship and travel to that distant land. I'm not sure when or if this will happen, but if it does, I hope my message will reach through time to the people who build that ship. I'm not asking them to believe me. I'm only asking them to consider whether my thoughts are plausible, because in the end, my request is simple: whoever you are, please don't forget to pack a net, some jars, some vials, and some plastic bags. These items won't take up much space, and they won't cost much, compared to the trip's expense. I just have the feeling that you might need them. I've found these simple things to be very useful on our Planet of the Bugs.

Acknowledgments

So, as fast as I could,
I went after my net.
And I said, "With my net
I can get them I bet.
I bet, with my net,
I can get those Things yet!"

DR. SEUSS, *The Cat in the Hat*

When I was 4 years old, my brother Ted gave me a homemade butter-fly net, and I've been fascinated with insects ever since. Some of my earliest memories are of chasing them in Detroit's Fargo Park, armed with my net, an empty glass jelly jar, and inspiration from Dr. Seuss to "get those things yet!" That's one of the great things about the science of entomology: you don't need much equipment to get started. Fifty years later, I'm a professor of entomology and curator of an insect museum. I've discovered and named 162 new insect species from 29 different countries all around the world. I've been fortunate to pursue a professional career centered on a lifelong interest, but I realize that it was possible only because many kind people supported and nurtured me along the way.

Writing this book took a lot longer than I expected when I started it. At several points I experienced writer's block and spent time pondering various parts. I'd like to express my sincere gratitude to Harold Greeney, founder of the Yanayacu Biological Research Station and Center for Creative Studies, and also I thank professors Lee Dyer of the University of Nevada at Reno and Tom Walla of Mesa State College for starting the research project that introduced me to Yanayacu, Ecuador. Yanayacu provided me with the biological inspiration and fresh air needed to finish this book.

Profound thanks to Professor Duncan Harris, director of the UW Honors Program, for encouraging me to teach three honors courses: Cosmology of Insects, Cosmology of Life, and Cloud Forest Ecology in Ecuador. Teaching these classes allowed me to sketch out the ideas presented in this book and gave me the opportunity to formulate them more clearly. I especially owe Professor Harris a debt of gratitude for allowing me to teach Cosmology of Insects, and not simply reminding me that such a discipline doesn't exist yet.

Research for this book was supported in part by a grant from the Wyoming NASA Space Grant Consortium, NASA grant #NGT-40102, and by Wyoming NASA EPSCoR NASA Grant #NCC5-578. My research grants from the National Science Foundation, Caterpillars and Parasitoids in the Eastern Andes of Ecuador, allowed me the opportunity to study, think, and write in the Andes highlands.

Over the years the ideas expressed here have been developed, discussed, and refined in several classes I've taught, especially Cosmology of Life, Insect Biology, Insect Classification, Insect Evolution, Aquatic Insects, Biodiversity, and Tropical Ecology. Hundreds of students have attended my lectures, and I thank them all for their patience, interest, and insights during this journey. In particular, I wish to thank Samin Dadelahi, Jen Donovan, Olivia Engkvist, and Nina Zitani for uniquely thoughtful discussions of my ideas, and, most importantly, for encouraging me to continue writing. Nina and Jen were the very first people to read the manuscript cover to cover, and the final book, although greatly metamorphosed from those earlier incarnations, is much improved by their attention to detail and enthusiasm for the subject.

My eldest brother, Ted, I thank for giving me my first insect net and loaning me his telescope. My father, Edward B. Shaw, I wish to thank for designing bigger and better insect nets and for collecting expeditions as numerous as the stars and insects. My mother, Vesta, I thank for allowing me to play with glass jars, poisonous chemicals, and especially for letting me bring live insects into the house. My brother, Tim, and my Uncle Lawrence (Latzy) McKay, I thank for introducing me to the genre of science fiction and inspiring me to bring science into the public domain.

When I started college at Michigan State University in 1973, my first declared major was astrophysics. I studied astronomy and learned about stellar evolution, the life and death of stars. But ultimately my childhood interest in insects won out, and I switched my major to entomology. Professor Roland Fisher was responsible for rekindling my interest in insects through his inspirational lectures, and he showed me what I wanted to do with my professional career. Professors Fred Stehr and Rich Merritt contributed immensely to fueling my passion for the study of insects. During my graduate studies, Professor Charles Mitter at the University of Maryland taught me not to be satisfied with the study of one discipline. While at Harvard University from 1984 to 1989, several individuals were influential. I wish to thank Professor Edward O. Wilson for inspiring me to study the social insects and tropical ecosystems and for deepening my appreciation for earth's biodiversity. He demonstrated the determination needed to be a successful writer and helped me through difficult times. He also provided me with inspirational advice on writing and publishing, without which I probably wouldn't have had the energy, direction, or determination to finish this manuscript. Professor Wilson is truly a gentleman and a scholar. Emeritus Professor Frank Carpenter was also inspirational and led me to study fossil insects. His generous contribution of fossil insect photographs allowed me to teach my first class on insect evolution. While at Harvard I also attended many lectures by Professor Stephen Jay Gould and read his popular natural history books. He profoundly influenced my worldview and my teaching style and demonstrated better than most how enthusiastically science can be related to the public.

From the University of Wyoming, I wish to thank Professors Greg K. Brown, Ron Canterna, Mark Lyford, and Terry Roark for providing knowledge, advice, books, information, comments, and encouragement. I also wish to thank the UW Berry Biodiversity Conservation Center, directed by Carlos Martinez del Rio, which contributed to the costs of the color plates. I'm very grateful to Dr. Danita Brandt (Invertebrate Paleontology, Department of Geological Sciences, Michigan State University) for her thoughtful technical and editorial notes on chapter 2 and for graciously allowing me to outline her research in that chapter. Dr. Conrad Labandeira (Smithsonian Institution) and Dr. Michael Engel (University of Kansas) both thoughtfully read chapter 5 and made many helpful suggestions. Dr. Douglas H. Erwin (Smithsonian Institution) graciously read chapter 6, which relies heavily on his pioneering research on that subject. Brandon Drake carefully proofread chapter 8 and contributed to my more accurate depictions of dinosaurs.

I wish to thank the following people for generously contributing photographs of living insects and other arthropods: Jennifer Donovan-Stump (Trinity School, New York, New York), Dr. Janice Edgerly-Rooks (Santa Clara University, California), Andy Kulikowski (Casper, Wyoming), Kevin Murphy (Irish Wildlife Photography, Westport, Ireland), Kenji Nishida (Monterverde, Costa Rica), Angela Ochsner (Torrington, Wyoming), David E. Rees (Timberline Aquatics, Fort Collins, Colorado), and Dr. Barbara Thorne (University of Maryland, College Park). Images of insects in amber were kindly contributed by Dr. Vincent Perrichot (Université de Rennes, France) and Dr. George Poinar Jr. (Corvallis, Oregon). The following people assisted with obtaining permission to publish images of insect fossils: Dr. Olivier Bethoux and Aurélie Roux (Muséum national d'Histoire naturelle, Paris, France) and Dr. Brian Farrell, Dr. Philip Perkins, Amie Jones, and Catherine Weisel (Museum of Comparative Zoology, Harvard University, Cambridge, Massachusetts). Marlene L. Carstens (University of Wyoming Photo Services) assisted with the production of digital images from black and white negatives. Helmuth Aguirre, my graduate assistant, kindly helped with the arrangement of images into plates.

I especially wish to thank Professor Paul E. Hanson at the University of Costa Rica for years of friendship and kind assistance with local arrangements for travel to La Selva Biological Research Station and San Ramon Biological Reserve, which inspired parts of this book. Paul, you really helped me "get my feet wet" in the tropics, for which I am very appreciative.

I owe the greatest debt of gratitude to my editor at the University of Chicago Press, Christopher Chung, for seeing merit in my writing and patiently working with me, providing thoughtful and detailed comments, and helping me to reshape my bulky manuscript into a sleeker and immensely more readable book. All my readers will benefit from Christopher's vision and hard work. Mary Gehl copyedited the final manuscript and made many insightful suggestions and corrections.

Stephen King I thank for his published thoughts about time travel and the craft of writing. Jane Auel I thank for her inspirational lecture at the University of Wyoming about her history as a writer. Many thoughtful people provided inspiration, helpful suggestions, information, and corrections; any errors that remain are my own.

I hope this book inspires the next generation of bug hunters with the same passion that my first butterfly net infused into me.

Finally, I wish to thank my wife, Marilyn, and my sons, Matthew and Michael. It's not easy living with an aging, distracted, and absent-minded entomology professor. Sometimes it's been hard to pay attention at the dinner table when I was trying to work out the history of the earth in the back of my mind. Your love and support through this process has been crucial, and I am profoundly grateful.

About the Author

Scott R. Shaw was born in Detroit, Michigan, in 1955. He started collecting insects at the age of 4. From 1973 to 1978 he attended Michigan State University where he studied astrophysics and entomology. He attended the University of Maryland from 1979 to 1984, where he obtained MS and PhD degrees in entomology. From 1984 to 1989, he worked at Harvard University in the Museum of Comparative Zoology. Since 1989, he has lived in Laramie, Wyoming, where he is professor of entomology at the University of Wyoming and Insect Museum curator. Professor Shaw has discovered and named 163 new insect species (mostly wasps) from 29 different countries. Fifteen insect species have been named after him by other scientists. He has published more than 114 scientific articles about insect classification and evolution. He has named insect genera (*Betelgeuse, Rigel, Orionis*) after stars in the sky and stars of late night television (*Marshiella lettermani*, a wasp named for David Letterman). His published suggestion for a Wyoming state insect, Sheridan's green hairstreak butterfly, was adopted by the Wyoming legislature and the governor in 2009. He has extensively studied insects in Costa Rica and Ecuador. This is his first book.

Notes

CHAPTER ONE

1. Many people are surprised to learn how much we still do not know about life on our own planet. We don't even know how many species we share the world with. Estimates range from seven million to a hundred million, and most biologists would agree that the vast majority of unknown species are insects living in the canopies of tropical forests.

2. E. O. Wilson, 1990. "First word," *Omni*, September, 6, 1990, *Academic Search Premier*, EBSCO*host* (accessed November 19, 2013).

3. David M. Raup, *Extinction: Bad Genes or Bad Luck?* (New York: W. W. Norton, 1991), 14.

CHAPTER THREE

1. However, expect them to live a long time. My colleague Nina Zitani's pet millipede lived for nearly twelve years.

2. The latest molecular studies of fungi suggest that the major lineages of fungal diversity evolved in tandem with the diversification of early vascular plants and terrestrial ecosystems. Fungal diversification certainly contributed to the evolution of microbial soils suitable for the colonization of land plants, and fungi also contributed to the diets of scavenging arthropods such as millipedes and symphylans. You can read more about fungal evolution in Robert Lücking, Sabine Huhndorf, Donald H. Pfister, Eimy Rivas Plata, and H. Thorsten Lumbsch et al., "Fungi Evolved Right on Track," *Mycologia* 101 (2009): 810–22.

CHAPTER FOUR

1. I use the term "stroll" only artistically here. Recent research on the 360-million-year-old amphibian, *Ichthyostega*, suggests that it just inched along by bending and straightening its back. The earliest amphibians probably dragged their hind legs and tail.

2. According to my colleague, engineering professor John McInroy, there is a fundamental reason to use six legs. With six legs it is possible to translate and rotate in all three directions. Also, six-legged creatures can resist forces and torques in all directions. The stability of six legs is well known in the field of robotics. For more

on walking robots, see Jean-Pierre Merlet, *Parallel Robots* (Dordrecht: Springer, 2005).

3. Perhaps I should point out that moving their legs three at a time is the way insects walk or run on a flat surface. Insects, because of their very small size, are also able to walk on vertical and even inverted surfaces. Recent research by S. N. Gorb indicates that they walk a bit more carefully on inverted surfaces. A fly on the ceiling moves slowly and carefully, leaving four legs planted and repositioning only two legs at a time. Upside-down walking on inverted surfaces is possible because insects are so small that the forces of surface tension and cohesion are proportionally greater. Attachment to both smooth and rough surfaces is improved by a variety of microscopic adaptations at the tip of the insect foot, including claws, hairy pads, and adhesive secretions. These characteristics were probably not present in the very earliest terrestrial hexapods, but were developed and refined in the lineages of flying insects over hundreds of millions of years. See S. N. Gorb, "Uncovering Insect Stickiness: Structure and Properties of Hairy Attachment Devices," *American Entomologist* 51 (2005): 31–35.

4. Although *Rhyniella praecursor* is the oldest undisputed hexapod, most entomologists do not consider the springtails to be true insects. They have very distinctive and unusual retracted mouthparts and appear to be a lineage of hexapods that diverged early from the line leading to most modern insects. However, older entomologists often called any six-legged arthropod, including springtails, an insect.

5. Until fairly recently, the archeognathans were combined with the silverfish and firebrats into a larger order called Thysanura, a name that is now being abandoned but that persists in some field guides and older literature. The extinct order Monura was described to include similar species that had only one bristlelike tail. Recent entomologists treat the monurans as members of the order Archaeognatha.

CHAPTER FIVE

1. This is an example of what ecologists call lekking behavior.

2. I'm taking artistic liberty with the mayfly story. While we are fairly certain that the mayflies, or at least the stem group of mayfly-like insects, first evolved in the Carboniferous years, we do not know for sure exactly when this line of insects first evolved freshwater, aquatic immature stages. They may not have done so until the Permian or even the Triassic years. But I'm guessing that they did go aquatic during the Carboniferous because of the age's abundant wetlands and the resource advantages of moving first into the freshwater niches. If freshwater fish existed then, it seems reasonable to assume there must have been aquatic insect naiads for them to consume.

3. Lepidosaurian reptiles were gliders during the Triassic period, with forms similar to today's Southeast Asian *Draco*. They may have been some of the earliest vertebrates to pursue and feed on winged insects in the air.

4. Cordaites plants had strap-shaped, broad-leaved foliage, as well as cones, and are considered to be closely related to the earliest conifers.

5. One diverse group of plant-decomposing arthropods was present in the Late Carboniferous: the oribatid mites. Primitive wingless insects and millipedes were also responsible for some of the decomposition.

6. While the beetles (order Coleoptera) have been known from Permian fossils for quite some time, only recently has evidence for Carboniferous beetles been discovered, by Béthoux (O. Béthoux, "The Earliest Beetle Identified," *Journal of Paleontology* 83 [2009]: 931–37). While the first beetles, as well as some other kinds of insects with complex metamorphosis, may have first originated during the Late Carboniferous, they were still rare and not diverse, and consequently did not yet have a profound ecological impact on forests. My discussion of beetles will await the next chapter, on the Permian period, when beetle species diversified and became common.

7. I do not mean to imply that coal production ceased after the Carboniferous, but only that from the period onward, the diversity of decomposing organisms, and competition for plant materials as food, increased. There are some significant coal deposits from much more recent times, such as the Paleocene open-pit mines near Gillette, Wyoming.

8. Along with the jumping bristletails, the silverfish were formerly placed in a larger order called Thysanura, a name that is still to be found in older literature. They are separated now because of their different mandible forms, as discussed in the last chapter.

9. Scientists have recently discovered that modern wingless insects, including silverfish and even worker ants, can manage a kind of gliding flight while falling from trees. Their bodies seem to function as an airfoil. During free fall they can turn and glide back to the tree trunk, and so avoid falling all the way to the ground.

10. Ancient paranotal lobes share a similar venation pattern with modern fully formed wings, which implies that wings evolved from appendages resembling the paranotal lobes.

11. This is clear because fossil paleodictyopteroid nymphs from Mason Creek are associated with *Macroneuropteris* foliage and have plant spores in their guts (Conrad Labandeira, personal communication).

12. We don't know the exact date of the very earliest flying insect, but we do know that winged insects were abundant and diverse by the Late Carboniferous, 320 million years ago. They may have maintained sole ownership of the airways for 100 to 150 million years or more before vertebrates finally took to the air.

13. Evidence suggests that the Paleodictyoptera were not the only plant-feeding insects that evolved during the Carboniferous. Other feeding styles, from fossil plant damage and coprolite evidence, include boring into plant tissues and stems, preying on seed fern seeds, and feeding (hole-feeding, margin-feeding, and scratching the surface) on the external foliage of multiple Carboniferous plant species.

14. In addition to spores, coprolites preserve microscopic fragments of plant tissues and wood which are assignable to particular plant-host species. So the study of coprolites provides insight into the herbivore consumption patterns and food webs of ancient times.

15. We have limited knowledge of the Carboniferous plants' chemical defenses. The common kinds of chemical defenses in modern plants, such as phenols, alkaloids, and tannins, degrade into by-products during fossilization, and cannot be recovered. However, fossils of some Carboniferous seed ferns contain resins, which preserve unique secondary chemical compounds that, according to Smithsonian paleontologist Conrad Labandeira, probably were used as insect-deterring compounds.

16. I'm making a bold assumption here. Fossil evidence as to where immature griffenflies lived is very rare. They might have been fully aquatic, like modern dragonflies, or they might have been terrestrial or semiaquatic in the humid forest undergrowth. We do know that the forests were filled with diverse potential predators, such as centipedes, spiders, scorpions, and amphibians. It seems to me that in order to grow as enormously large as an adult griffenfly, the developing young must have lived in a somewhat sheltered habitat. At least they would have avoided spiders, centipedes, and scorpions in the fresh water ponds, so developing there seems far more likely.

17. Because some insect paleontologists think that the Carboniferous roaches are considerably different from modern ones, they refer to these creatures as "Paleozoic roachoids" or "ovipositor-bearing cockroaches." This terminology is cumbersome, so I've elected to use more familiar terms and simply call all of the ancient species "roaches" or "cockroaches." Although they are unlike modern roaches in some important ways (for instance, they have a visible ovipositor), if we could travel back to the Carboniferous, we would readily recognize them as roaches and probably declare the place one huge roach-infested swamp.

CHAPTER SIX

1. Fans of movie trivia may recall that Clint Eastwood made his brief debut in this film as the heroic jet fighter pilot who fires the rocket that destroys the giant mutant spider.

2. This developing drama of the warm-bloods has one interesting side note. About this time some small insects of the now-extinct order Diaphanopterodea evolved long, slender mouthparts, and some of them resembled mosquitoes. This is the first instance of possible blood feeding by insects, perhaps not coincidentally at the same time that warm-bloodedness appeared. Maybe insects were going on the attack. I'm just speculating, but did ancient blood-sucking insects spread diseases among the herds of protomammals? Living mosquitoes are known to transmit more than two hundred kinds of blood-borne diseases, so it is certainly plausible that Permian blood-feeding insects, such as diaphanopteroids, might have transmitted fatal diseases among the herds of Tartarian protomammals.

3. Order diversity was greater then because the Permian was a combination of lingering Carboniferous lineages, in addition to numerous later lineages, that diversified at that time and had descendants that survived the extinction. These survivors formed the core of the modern insect fauna that thrives today.

4. As we discussed in chapter 2, trilobites originated and were most diverse during the Cambrian period. They become scarcer, relative to other marine groups, in Paleozoic-era sediments. Even so, I selected trilobites as my symbol of the entire era because they persisted across that time, but do not appear at all in Mesozoic or Cenozoic sediments. For the end of the Permian, fossils of brachiopods, bryozoans, and crinoids are better geological markers because they were more common then. By the Late Paleozoic, trilobite species diversity was very low, so perhaps their extinction was inevitable.

5. Volcanic explosions across Permian Siberia spewed 1.5 million times as many airborne particles as the 1981 eruption of Mount Saint Helens.

6. Fossils of Permian grylloblattids still show wings, but by the Late Cretaceous period they were wingless.

7. Some authors treat the homopterans as part of the order Hemiptera (discussed in the next chapter), and divide them into three suborders: Sternorrhyncha, Auchenorrhncha, and Coleorrhyncha. This nomenclature is obviously a bit cumbersome, so I prefer to refer to these insects by their simpler and more familiar name: homopterans.

8. The oldest fossil insect larva and the oldest fossils of adults from several groups known to have complex metamorphosis are from the Permian period. There are some fossil plant galls from the Late Carboniferous, and since modern galls are mostly caused by insect larvae, some scientists, such as Smithsonian paleontologist Conrad Labandiera, have speculated that complex metamorphosis first developed late in the Carboniferous. The discovery of some putative holometabolous insects from the Late Carboniferous led Nel and colleagues to refer to complex metamorphosis as "a crucial innovation with delayed success" (A. Nel et al., "The Earliest Holometabolous Insect from the Carboniferous: A 'Crucial' Innovation with Delayed Success [Insecta: Protomeropina: Protomeropidae]," *Annales de la Société Entomologique de France*, n.s. 43 [2007]: 349). Most entomologists credit the drier climate (Permian aridity) for stimulating the diversification of holometabolous insects during the Permian period. Whatever the reason, we do know that they diversified explosively during the Late Permian, and that the groups with complex metamorphosis survived the end Permian well.

9. Nel and colleagues have suggested that scorpionflies may have originated during the Late Carboniferous, but they become common as fossils during the Permian (Nel et al., "The Earliest Holometabolous Insect from the Carboniferous").

10. Sometime during the middle Mesozoic era, a lineage of scorpionflies diverged and evolved into blood-feeding ectoparasitic parasites of birds, mammals, and possibly dinosaurs. The order Siphonaptera, commonly called fleas, are now known to be most closely related to the snow scorpionflies, family Boreidae. The fleas evolved rapidly along with mammals during the Cenozoic era, and now comprise at least 2,500 living species, so the modern Mecoptera are substantially more species-rich if the fleas are reclassified and considered as part of the scorpionflies.

11. R. J. Mackay and Glenn B. Wiggins, "Ecological Diversity in Trichoptera," *Annual Review of Entomology* 24 (1979): 185.

12. Nel and colleagues have suggested that the stem group of these orders may have originated as far back as the Pennsylvanian subperiod of the Carboniferous. But the caddisflies did not diversify until the Permian, and moths only much later, during the Late Mesozoic.

13. Douglas H. Chadwick and Mark W. Moffatt, "Planet of the Beetles," *National Geographic* 193 (no. 3): 100.

14. Although inconspicuous, the Psocoptera have diversified greatly and there are at least 4,400 described species. Since they are tiny and live in concealed places, there probably are many bark lice species still undiscovered.

15. This isn't the first time we have considered continental drift's profound impact on the history of life. Back in chapter 2 we discussed how late Precambrian continental drift aligned the continents in a way that led to the global Varanger ice ages, which probably caused massive extinctions among ancient microbial life. We also considered how ongoing continental drift brought the planet out of the Varanger ice ages, possibly triggering the Cambrian explosion of life.

CHAPTER SEVEN

1. Along with our Thanksgiving turkey, all modern birds and certain kinds of dinosaurs, including the feathered ones, have a wishbone. This bone is an example of what evolutionary biologists call a synapomorphy—a uniquely shared characteristic that provides evidence of common ancestry. You can think of it as another time message if you wish.

2. Although *Apatosaurus* is now the correct scientific name for *Brontosaurus*, I take the artistic liberty of using "brontosaur" as the common name, as it is more easily recognized. This is not the first time that a disused scientific name was adopted for a common one: you have probably heard of the duck-billed *Platypus*, but you may not know that *Platypus* is no longer standard; instead, this animal is now called *Ornithorhynchus anatinus*.

3. Until the end of the Cretaceous, our mammalian ancestors were little more than tiny and furry shrewlike insectivores that scampered through the forest leaf litter, hid under logs, and no doubt lived in constant fear of the massive predatory dinosaurs. It was a dreadful time in our history for sure, but the tasty and nutritious insects sustained us mammals until the K-T asteroid finished off the last of the big, nasty brutes. But now with the discovery that birds are dinosaurs, too, we have to acknowledge that the dinosaurian dynasty that ruled the Mesozoic has also done well for itself in recent times.

4. Our particular culture doesn't care to eat insects all that much, but we still can't avoid them. Most of us consume about a pound or more each year, ground up in flour or cereals or mixed in with fruits and vegetables, and some studies have suggested that our cultural aversion to eating bugs may be contributing to B vitamin deficiencies. Some people would like to exclude all insect parts from our diets, but there's a couple of good reasons why the Food and Drug Administration has been unable (or unwilling) to do that. First, it's virtually impossible to screen all

the insects from plant food sources. Second, there's no real good reason to keep them out because, in most cases, including insects with plant materials actually improves the nutritional value of our food.

5. I'll have much more to say about wasp diversity in the following chapters. For now, it should suffice to tell you that wasps are one of the hyperdiverse insect orders, with species diversity perhaps comparable to the beetles, and that the most diverse lineages of wasps have evolved parasitic behaviors.

CHAPTER EIGHT

1. Goliath beetle adults may be among the heaviest insects that ever lived, but we have only recently discovered that the immature forms of some South American rhinoceros beetles are even heavier. The larva of one of these, the Hercules beetle, is known to weigh as much as 120 grams. Many insects reach their peak body weight not as adults but in their last larval stage, just prior to pupating; since the immature forms of large tropical beetles feed deep in wood, it is very possible that we have not yet determined the heaviest living insect.

2. Holland's description of the new dinosaur is included in a longer paper about the characteristics of its head: W. J. Holland, "The Skull of *Diplodocus*," *Memoirs of the Carnegie Museum* 9 (1927): 379–403. Although from 1900 to 1931 Holland published fifty-five papers about dinosaurs and named and described two new species, he is much better known for his entomological works. Over the course of his professional career he published some five hundred scientific papers, mostly about insects. His best known works certainly were *The Butterfly Book*, published in 1898, and *The Moth Book*, published in 1903. No other books of the time were so influential in introducing so many people to the study of butterflies and moths.

3. However, the *Utahraptor*, which was found in Utah after the book and first movie appeared, was about the size of the *Velociraptor* depicted in the movies.

4. In the end, we don't know for sure if the allosaurs hunted in packs, but we do know that they fed on meat, and that it was a dangerous business. On display in the University of Wyoming Geology Museum is a cast of one of the most complete juvenile *Allosaurus* skeletons ever unearthed. Affectionately dubbed Big Al, the skeleton is notable not so much for its completeness but more for the imperfections of some bones. The gnarly outgrowths on several ribs and one foot provide solid evidence of past wounds partly healed. Allosaurs fought hard for their meals, and they sustained serious injuries in the process.

5. We don't know the actual number of living wasp species, but it is enormous. The reason is simply that many of them are microscopic, undiscovered, and unnamed. Hymenopterists, the entomologists who study wasps, have already named around a hundred thousand species, and most believe that there are millions. Many think that the species diversity of microscopic parasitic wasps is comparable to that of the hyperdiverse beetles. Some (myself included) think that wasp diversity may actually exceed that of beetles.

6. Primitively the hymenopteran ovipositor was composed of four shafts; how-

ever, in most living wasps the upper two are fused into one inverted U-shaped upper valve. Exceptions include some sawflies where the ovipositor tip is divided into four shafts, possibly a remnant of the primitive condition, and some ophioniform Ichneumonidae, which have the upper shafts divided except at the tip (thought to be a secondarily evolved condition). In most modern wasps there are three flexible shafts: a broader upper shaft, and two narrower lower shafts.

7. These wasp-associated microorganisms are what we call the polydnaviruses, and they appear to be one of the key reasons for the endoparasitic insects' vast success.

8. These extreme specializations of body form across the larval lifespan are an example of hypermetamorphic adaptations, or hypermetamorphosis.

9. The sudden development of silk glands in emerging parasitic wasp larvae is pretty impressive when you consider that the vast majority of other young silk-spinning insects, like caterpillars, develop these glands over their entire larval lifetime. Wasp larvae don't need silk at all while they are feeding inside a host insect, so they repress their gland's development until the final molt.

10. Historically, termites have been classified as a distinct insect order, Isoptera, but recent studies treat them as a lineage within the order Blattaria, or Blattodea, the cockroaches.

11. The idea that termites are social feeding colonies bound together by their common need to exchange gut symbionts is known as the symbiont hypothesis of social evolution. To be fair, this is another controversial topic, and several other viable hypotheses may explain the origin of the termite's social behavior. One popular idea is that since termites develop slowly and live in hidden habitats rich in concentrated food, they enjoy several advantages to staying within their multigenerational family groups. They mutually benefit from sharing food, more easily defending the group, and jointly caring for the young. It is very risky for individuals to disperse and find suitable locations for new successful nests; most attempts at founding new colonies certainly end in failure. Ultimately, individuals that stay inside an established colony have a much better chance of surviving.

12. *Archaeopteryx* has long been regarded as the oldest bird until the recent discovery of the Chinese feathered dinosaur *Microraptor*, which had apparent flight feathers on both its front and hind legs. The discovery of *Microraptor* has rekindled the hot debates about the origin of birds and bird flight. I don't expect to resolve them here. Rather, I'd just like to observe that even if *Microraptor* was fully arboreal, it must have descended from formerly ground-dwelling ancestors. All of the little raptors were toothy carnivores, so they certainly included insects in their broad diet. Whether *Archaeopteryx* or *Microraptor* is the subject of our discussion, it is worth noting that chasing insects into the trees and air likely drove the origin of bird flight.

13. Robert Nudds at the University of Manchester in England has recently studied the load-carrying capacities of fossil feathers from *Archaeopteryx* and the ancient Chinese bird *Confuciusornis*. He found that these early birds had flimsy

feathers and probably did not fly very well. They may have just glided from branch to branch, or used their wings to slow their descent when falling. Nudds's research was profiled in *Science News*, June 5, 2010.

14. *Archaeopteryx* is usually pictured as nibbling on a little lizard or a fish. I don't doubt that they ate lizards, salamanders, and such, but certainly, like their ancestors, they were highly omnivorous and widely insectivorous.

15. This is just the total for the bird lice. The Phthiraptera subsequently radiated extensively on mammalian hosts during the Cenozoic era, so the total of living bird and mammal lice is nearly five thousand species.

CHAPTER NINE

1. The Andes Mountains of western South America were formed over the last 140 million years as the continent was pushed westward and into the Pacific crust by the widening of the South Atlantic's ocean floor. This was a long, slow process, and the western parts of South America did not become mountainous too quickly. The main uplifting of the Andes occurred more recently, between 23 million and 5 million years ago. The Andes were not tall enough to direct the Amazon River eastward across its present pathway to the Atlantic until about 10 million years ago. Likewise, the Himalayas did not form until comparatively recently, during the last 60 million years, as India collided with the Asian continent.

2. The Gnetales are examples of dioecious plants: their sexual organs are located on separate individuals.

3. The Lepidoptera share with their sister group, Trichoptera (caddisflies), the same silk-producing mechanism that we discussed in chapter 6. Even as tiny leaf miners, primitive caterpillars were lining their tunnels with silk secreted from modified salivary glands. Silk—even those fine silkworm strands that we cherish as ties, shirts, scarves, stockings, and underwear—is little more than dried insect spittle.

4. These kinds of defensive chemicals are restricted to particular plants, and occur in varying amounts among them. Since they are not required for the primary metabolism of plant growth, scientists call them secondary chemicals. Although some of these chemicals may have originated from developmental waste products, many seemingly evolved strictly to defend against the insects' plant-feeding behaviors.

5. Perhaps this doesn't do enough justice to the tunneling abilities of some bees, which nest in a wide variety of soil types, including both soft soils and heavy clays. Impressively, some Wyoming bees are known to tunnel nests into rock—into the soft sandstone of the Laramie Formation.

6. You may recall that when we first considered the parasitoid wasps' early evolution, we discussed the importance of the larvae's closed hind gut, which prevents the young from fouling their local feeding area with their own feces. This adaptation was equally helpful to the nest-provisioning solitary wasps, and ultimately to

social bees, ants, and paper wasps. Like termites, these groups needed to solve the problem of sewage control before becoming abundantly social. While the termites consume their feces, the wasps hold it until after their larval feeding is complete.

7. One exception is the weaver ants, which stimulate their larvae to spin silk by squeezing them. These ants then tie together leaves with the silk strands and form nests in trees.

8. While the kin selection hypothesis has dominated the literature for several decades, a recent controversial paper by E. O. Wilson and colleagues challenges it and refocuses attention on other ideas, such as the mutual benefits of nest sharing for food gathering and defense. For more information, see Martin A. Nowak, Corina E. Tarnita, and Edward O. Wilson, "The Evolution of Eusociality," *Nature* 466 (2010): 1057–62.

9. If you still doubt whether or not big dinosaurs could feel a wasp's sting, please consider this. An insect sting's painfulness is categorized on a scale of 0 to 4 by the Schmidt Sting Pain Index, which was created by the eminent entomologist and wasp specialist, Dr. Justin Schmidt. The sting of the most painful insect, the South American bullet ant, is described as "pure, intense, brilliant pain, like firewalking over flaming charcoal with a 3-inch rusty nail grinding into your heel." Justin O. Schmidt, "Hymenoptera Venoms: Striving toward the Ultimate Defense against Vertebrates," in *Insect Defenses: Adaptive Mechanisms and Strategies of Prey and Predators*, ed. D. L. Evans and J. O. Schmidt, 387–419 (Albany: State University of New York Press, 1990).

10. For a good synopsis of the multiple dinosaur extinction theories, see Robert Bakker's book *The Dinosaur Heresies*. Bakker himself favors the hypothesis that, during the late Cretaceous, dinosaurs were naturally declining because the Asian and North American populations mixed after these areas were connected by the Bering Land Bridge. Many species may have gone extinct due to competition over limited resources or because of new invading predators well before the asteroid finished of the last of the big dinosaurs.

11. Recent studies of the Deccan Traps lava beds in India have brought volcanoes back into the limelight. These volcanic eruptions have been dated near to the end-Cretaceous mass extinction, and were massive enough to have caused global climate change and possibly some extinction, especially in marine ecosystems. This doesn't refute the importance of an end-Cretaceous asteroid impact, but it suggests that many species may have already been declining before then.

12. Nevertheless, the asteroid-impact hypothesis draws attention to the fact that big rocks occasionally collide with this planet and may affect patterns of life over the long haul. Some paleontologists plotted the data for species diversity over time and noticed that the extinctions at the end of the Cretaceous and the end of the Permian are not the only major ones—they are just two of the largest. Mass extinctions appear to be periodic and repeat on a somewhat cyclical basis. On average, it looks as if an extinction event of varying intensity occurs about once every twenty-six million years. Granted, this could just be a coincidence, but it prompted scientists to speculate about how and why such events might regularly repeat. The

dominant notion is that the earth is getting hit by asteroids, and that something out there with gravitational pull is affecting their stable orbits, as well as the orbits of comets, on a cyclic basis, about once every twenty-six million years. Perhaps either a remote small planet or dark star with a broad twenty-six-million-year orbit perturbs them whenever it passes near to our solar system.

CHAPTER TEN

1. The oldest fossil hominid is *Ardipethecus ramidus*, found in the Afar Rift of Ethiopia and estimated to be 4.4 million years old. Its skeletal anatomy suggests that it lived in trees, had a largely plant-based diet, and possibly had bipedal capability. There were definitely bipedal hominids in eastern Africa by 3.6 million years ago, demonstrated by the Laetoli fossil footprints in Tanzania, left in volcanic ash by two individuals. *Australopithecus afarensis*, a short, 3.5-foot-tall hominid known from the famous Lucy skeleton, evolved by 3.2 million years ago. By 2.5 million years ago there were abundant East African hominids in the form of *Australopithecus africanus*, followed by *Australopithecus boisei* around 2 million years ago. The later species had flat molar teeth assumed to be well adapted for grinding plant foods. Shortly after this time, *Homo rudolfensis* evolved, which is considered to be the most ancient species in our genus *Homo*. By 1.5 million years ago, *Homo erectus* had arrived, and the migration out of Africa is thought to have begun. By about half a million years ago, *Homo sapiens* and Neanderthals evolved and were coexisting.

2. Our philosophy of science relies on parsimony, or simplicity, for selecting a preferred hypothesis. If chimps and humans use tools, the most parsimonious explanation is that both inherited the behavior from our common ancestor. This makes plenty of sense when you consider that chimps and homonids evolved from insectivorous tree-dwelling primates. A more complicated—and less likely—explanation is that tool use evolved independently in both groups.

3. If big dinosaurs had survived the end-Cretaceous asteroid impact, and they, instead of us, had evolved big brains and consciousness (and dinosaur egos), perhaps they would call the last sixty-five million years the Age of Dinosaur Civilization.

4. This simple method of panning for insect gold was pioneered and perfected by the renowned entomologist Lubomir Masner, and it works remarkably well for sampling micro-Hymenoptera (microscopic wasps).

5. Since I wrote this chapter, I have returned to Ecuador five more times and continued to work on my elusive new wasp. My description of the new genus and species was published during May 2012 in the *International Journal of Tropical Insect Research*, so the organism now has a formal scientific name: *Napo townsendi*, the genus named after the province of Ecuador where it lives, and the species named after Andrew Townsend, my student who first collected the wasp. In May 2012, by lucky coincidence I happened to be in Ecuador at the same time my research paper was published, and I had the exciting experience of being the first person to see a living, newly named *Napo townsendi*. I should mention that in June 2010, my col-

league Dr. Nina Zitani was the first to spot living, although unnamed, specimens perching on some trailside leaves.

We have discovered that the males, while still scarce, can predictably be found perching on leaves of *Dendrophorbium* tree seedlings. I've hypothesized that the male wasps are using these select leaves as platforms for attracting females. During May 2012 and 2013, along with my colleague Dr. Will Robinson from Casper College and our student research assistants, Delina Barbosa and Andy Kulikowski, I learned that the male *Napo* wasps demonstrate a sequence of predictable behaviors that suggest they might be using sex pheromones to mark leaves and attract distant, rare females. Will discovered that the males show remarkable fidelity to particular leaves, often spending several days on the same leaf or returning to the same one after briefly flying away. On sunny mornings, the males groom themselves, first rubbing their hind legs over the tips of their abdomens then over their wings' upper and lower surfaces. When the males are at rest, early in the morning or late in the afternoon, they normally fold their wings over their bodies, but immediately after grooming, they hold their wings outward at 45° angles. We've hypothesized that this wing posture is a calling behavior that ventilates pheromones rubbed onto the wing surface. Later in the summer of 2012, I sent a preserved male specimen to Dr. Donald Quicke, a world expert on the glands of parasitic wasps, at Imperial College London. He dissected the abdomen and documented that there is indeed a unique set of glands along its back sides, near the tip, exactly where we predicted.

Female *Napo* wasps remain comparatively rare and elusive. In four years, we have briefly spotted only ten, while they were mating. We are continuing to study *Napo townsendi*, documenting them with photographs and video, but I think we are making good progress in understanding how these rare and widely dispersed tiny insects manage to find each other in the forest.

6. Imagobionts are so named because their adult host insects are technically known as the imago stage. Based on amber fossils, we know that they had evolved by at least fifty million years ago, making their origin and diversification one of the Cenozoic era's notable events.

7. At our current pace, it could take another five hundred years just to identify, name, and catalog all the other living species. Clearly we don't have the luxury of taking that long. If we can count the digits of the number π to more than one billion decimal places, why can't we get on with the task of naming the several million species that reside with us? If we don't, we'll have no way to communicate information about these organisms and no way to associate new information with the same creatures.

8. You don't need to travel that far. If you are patient and spend a lot of time looking, you can find new microscopic insects living in your own back yard. I have found them in mine.

POSTSCRIPT

1. We have already devised remote-sensing space probes able to discover the conditions suitable for life. One of these probes, the *Galileo* spacecraft, made a pass near earth before heading to its final destination—Jupiter—and its instruments were directed toward our planet. From a distant vantage point in outer space, through spectral analysis and with infrared photography, we were able to detect three clear indications of life on earth. The first, the presence of an oxygen-rich atmosphere, is attributed entirely to the presence of photosynthetic life, for we don't know of any nonliving processes that would allow oxygen to accumulate to such exceptional levels (near 21 percent of the earth's atmosphere). It also hints at the possibility of oxygen-respiring animals, which stabilize oxygen production (by consumption) near those levels. The second, widespread light-absorption in a particular wavelength, is attributed to only one substance that we know of: the pigment chlorophyll in plants. The third, trace amounts of atmospheric methane, is particularly interesting because this gas is unstable in the presence of oxygen: it oxidizes into carbon dioxide and water. Its persistence indicates steady methane production by living organisms: by bacteria, social insects, and, lately, humans.

2. Haldane is quoted as saying this in Evelyn G. Hutchinson, "Homage to Santa Rosalia, or Why Are There So Many Kinds of Animals?" *American Naturalist* 93 (1959): 146. However, Hutchinson refers to it as "a story, possibly apocryphal." Haldane, in his book *What is Life?*, made a similar statement: "The Creator would appear as endowed with a passion for stars, on the one hand, and for beetles on the other" (John Burdon Sanderson Haldane, *What is Life? The Layman's View of Nature* [London: L. Drummond, 1949], 258).

Suggested Reading

CHAPTER 1: THE BUGGY PLANET

Berenbaum, May R. *Bugs in the System: Insects and Their Impact on Human Affairs.* Reading, MA: Addison-Wesley, 1995.
Marshall, Stephen A. *Insects, Their Natural History and Diversity.* Buffalo, NY: Firefly Books, 2006.
Moffett, Mark W. *The High Frontier: Exploring the Tropical Rainforest Canopy.* Cambridge, MA: Harvard University Press, 1993.
Wilson, Edward O. *The Diversity of Life.* New York: W. W. Norton, 1992.
Wilson, Edward O., and Frances M. Peter, eds. *Biodiversity.* Washington, DC: National Academy Press, 1988.

CHAPTER 2: RISE OF THE ARTHROPODS

Barnes, Robert D. *Invertebrate Zoology.* Philadelphia: W. B. Saunders, 1974.
Brandt, Danita S. "Ecdysial Efficiency and Evolutionary Efficacy among Marine Arthropods: Implications for Trilobite Survivorship," *Alcheringa* 26 (2002): 399–421.
Brandt, Danita S., D. L. Meyer, and P. B. Lask. "*Isotelus* (Trilobita) 'Hunting Burrow' from Upper Ordovician Strata, Ohio," *Journal of Paleontology* 69 (1995): 1079–83.
Erwin, Douglas, James Valentine, and D. Jablonski. "The Origin of Animal Body Plans," *Scientific American* 85 (1997): 126–37.
Gore, R., and O. L. Mazzatenta. "Explosion of Life: The Cambrian Period," *National Geographic* 184 (1993): 120–36.
Gradstein, F. M., and J. G. Ogg. "Geologic Time Scale 2004—Why, How, and Where Next!" *Lethaia* 37 (2004): 175–81.
Hoffman, P. F., A. J. Kaufamn, G. P. Halverson, and D. P. Schrag. "A Neoproterozoic Snowball Earth," *Science* 281 (1998): 1342–46.
Lageson, David R., and Darwin R. Spearing. *Roadside Geology of Wyoming.* Missoula, MT: Mountain Press, 1988.
Lane, Nick. *Oxygen, the Molecule That Made the World.* Oxford: Oxford University Press, 2002.
Lipps, J. H., and P. W. Signor, eds. *Origin and Early Evolution of the Metazoa.* New York: Plenum Press, 1992.

CHAPTER 3: SILURIAN LANDFALL

Beerbower, J. R., J. A. Boy, W. A. DiMichele, R. A. Gastaldo, R. Hook, N. Hotton III, T. L. Phillips, S. E. Scheckler, and W. A. Shear. "Paleozoic Terrestrial Ecosystems." In *Terrestrial Ecosystems through Time*, edited by A. K. Behrensmeyer, J. D. Damuth, W. A. DiMichele, R. Potts, H. D. Sues, and S. L. Wing, 205-35. Chicago: University of Chicago Press, 1992.

Gensel, P. G., and H. N. Andrews. "The Evolution of Early Land Plants." *American Scientist* 75 (1987): 478-89.

Gensel, P. G., and D. Edwards. *Plants Invade the Land*. New York: Columbia University Press, 2001.

Jeram, A. J., P. A. Selden, and D. Edwards. "Land Animals in the Silurian: Arachnids and Myriapods from Shropshire, England." *Science* 250 (1990): 658-61.

Lucking, R., S. Huhndorf, D. H. Pfister, E. R. Plata, and H. T. Lumbscyh. "Fungi Evolved Right on Track." *Mycologia* 101 (2009): 810-22.

Wellman, C. H., P. L. Osterloff, and U. Mohiuddin. "Fragments of the Earliest Land Plants." *Nature* 425 (2003): 282-85.

CHAPTER 4: SIX FEET UNDER THE MOSS

Cressler, W. L. III. "Evidence of the Earliest Known Wildfires." *Palaeos* 16 (2001): 171-74.

Edwards, D., P. A. Seldon, J. B. Richardson, and L. Axe. "Coprolites as Evidence for Plant-Animal Interactions in Siluro-Devonian Terrestrial Ecosystems." *Nature* 377 (1995): 329-31.

Engel, Michael S., and David A. Grimaldi. "New Light Shed on the Oldest Insect." *Nature* 427 (2004): 627-30.

Gaunt, M. W., and M. A. Miles. "An Insect Molecular Clock Dates the Origin of Insects and Accords with Paleontological and Biogeographic Landmarks." *Molecular Biology and Evolution* 19 (2002): 748-61.

Gensel, P. G., and H. N. Andrews. *Plant Life in the Devonian*. New York: Praeger Press, 1984.

Gorb, S. N. "Uncovering Insect Stickiness: Structure and Properties of Hairy Attachment Devices." *American Entomologist* 51 (2005): 31-35.

Greenslade, Penelope. "Collembola (Springtails)." in *The Insects of Australia*, vol. 1, edited by Ian D. Naumann, 252-64. Carlton: Melbourne University Press, 1991.

Grimaldi, David, and Michael S. Engel. *Evolution of the Insects*. New York: Cambridge University Press, 2005.

Labandeira, Conrad C., B. S. Beall, and F. M. Hueber. "Early Insect Diversification: Evidence from a Lower Devonian Bristletail from Quebec." *Science* 242 (1988): 913-16.

Rasnitsyn, Alexandr, and Donald L. J. Quicke. *History of the Insects*. Dordrecht: Kluwer Academic Publishers, 2010.

Retallack, G. J. "Early Forest Soils and Their Role in Devonian Global Change." *Science* 276 (1997): 583–85.

Rice, C. M. "A Devonian Auriferous Hot Spring System, Rhynie, Scotland." *Journal of the Geological Society* 152 (1995): 229–50.

Shear, W. A. "Early Land Animals in North America: Evidence from Devonian Age Arthropods from Gilboa, New York." *Science* 224 (1984): 492–94.

Westenberg, K., and J. Blair. "The Rise of Life on Earth: From Fins to Feet." *National Geographic* 195 (1999): 114–27.

Wheeler, Ward C., Michael Whiting, Quentin D. Wheeler, and James M. Carpenter. "The Phylogeny of Extant Hexapod Orders." *Cladistics* 17 (2001): 113–69.

CHAPTER 5: DANCING ON AIR

Berner, R. A., and D. E. Canfield. "A New Model for Atmospheric Oxygen over Phanerozoic Time." *American Journal of Science* 289 (1989): 333–61.

Béthoux, O. "The Earliest Beetle Identified." *Journal of Paleontology* 83 (2009): 931–37.

Brauckmann, C., B. Brauckmann, and E. Gröning. "The Stratigraphic Position of the Oldest Known Pterygota (Insecta, Carboniferous, Namurian)." *Annals of the Geological Society of Belgium* 117 (1996): 47–56.

Carpenter, Frank M. "Adaptations among Paleozoic Insects." *Proceedings of the First North American Paleontological Convention (1969)* 1 (1971): 1236–51.

Carpenter, Frank M. "Arthropoda: Superclass Hexapoda." In *Treatise on Invertebrate Paleontology*, edited by R. L. Kaesler, 1–655. Boulder, CO: Geological Society of America, 1992.

Djernaes, Marie, Klaus-Dieter Klass, Mike D. Picker, and Jakob Damgaard. "Phylogeny of Cockroaches (Insecta, Dictyoptera, Blattodea), with Placement of Aberrant Taxa and Exploration of Out-Group Sampling." *Systematic Entomology* 37 (2012): 65–83.

Dudley, R. *The Biomechanics of Insect Flight: Form, Function, and Evolution*. Princeton, NJ: Princeton University Press, 2000.

Hasenfuss, I. "A Possible Evolutionary Pathway to Insect Flight Starting from Lepismatid Organization." *Journal of Zoological Systematics and Evolutionary Research* 40 (2002): 65–81.

Kukalová-Peck, J. "New Carboniferous Diplura, Monura, and Thysanura, the Hexapod Ground Plan, and the Role of Thoracic Lobes in the Origin of Wings (Insecta)." *Canadian Journal of Zoology* 65 (1987): 2327–45.

Labandiera, Conrad C. "Paleobiology of Predators, Parasitoids, and Parasites: Accommodation and Death in the Fossil Record of Terrestrial Invertebrates." in "The Fossil Record of Predation," special issue edited by M. Kowalewski and P. H. Kelly, 211–50. *Paleontological Society Special Papers* 8 (2002).

Labandiera, Conrad C., and T. L. Phillips. "A Carboniferous Petiole Gall: Insight into the Early Ecological History of the Holometabola." *Proceedings of the National Academy of Sciences U S A* 93 (1996): 8470–77.

Marden, J. H., and M. G. Kramer. "Surface-Skiming Stoneflies: A Possible Intermediate Stage in Insect Flight Evolution." *Science* 266 (1994): 427–30.

McKittrick, F. A. "Evolutionary Study of Cockroaches." *Cornell University Agricultural Experiment Station Memoirs* 389 (1964): 1–197.

Nagamitsu, T., and T. Inoue. "Cockroach Pollination and Breeding System of *Uvaria elmeri* (Annonaceae) in a Lowland Mixed-Dipterocarp Forest in Sarawak." *American Journal of Botany* 84 (1997): 208–13.

Nel, A., P. Rocques, P. Nel, J. Prokop, and J. S. Steyer. "The Earliest Holometabolous Insect from the Carboniferous: A "Crucial" Innovation with Delayed Success (Insecta: Protomeropina: Protomeropidae)." *Annales de la Société Entomologique de France*, n.s. 43 (2007): 349–55.

Niwa, N., A. Akimoto-Kato, T. Niimi, K. Tojo, R. Machida, and S. Hayashi. "Evolutionary Origin of the Insect Wing via Integration of Two Developmental Modules." *Evolution and Development* 12 (2010): 168–76.

Rasnitsyn, A. P., and D. L. J. Quicke. *History of Insects*. Dordrecht: Kluwer, 2002.

Shear, W. A., and J. Kukalová-Peck. "The Ecology of Paleozoic Terrestrial Arthropods: The Fossil Evidence." *Canadian Journal of Zoology* 68 (1990): 1807–34.

CHAPTER 6: PALEOZOIC HOLOCAUST

Chadwick, D. H., and Mark W. Moffett. "Planet of the Beetles." *National Geographic* 193 (1998): 100–18.

Dalziel, I. W. D. "Earth Before Pangea." *Scientific American* 272 (1995): 58–63.

Erwin, Douglas H. *Extinction: How Life on Earth Nearly Ended 250 Million Years Ago.* Princeton, NJ: Princeton University Press, 2006.

———. *The Great Paleozoic Crisis: Life and Death in the Permian.* New York: Columbia University Press, 1993.

———. "The Mother of Mass Extinctions." *Scientific American* 273 (1996): 72–78.

———. "The Permo-Triassic Extinction." *Nature* 367 (1994): 231–36.

Hynes, H. B. N. "The Ecology of Stream Insects." *Annual Review of Entomology* 15 (1970): 25–42.

Jin, Y. G., Y. Wang, W. Wang, Q. H. Shang, C. Q. Cao, and Douglas H. Erwin. "Pattern of Marine Mass Extinction Near the Permian-Triassic Boundary in South China." *Science* 289 (2000): 432–36.

Knoll, A. H., R. K. Bambach, D. E. Canfield, and J. P. Grotzinger. "Comparative Earth History and Late Permian Mass Extinction." *Science* 273 (1996): 452–57.

Labandeira, Conrad. "Insect Mouthparts: Ascertaining the Paleobiology of Insect Feeding Strategies." *Annual Review of Ecology and Systematics* 28 (1997): 153–93.

Mackay, R. J., and Glenn B. Wiggins. "Ecological Diversity in Trichoptera." *Annual Review of Entomology* 24 (1979): 185–208.

Raup, David M. *Extinction: Bad Genes or Bad Luck?* New York: W. W. Norton, 1991.

Ward, P. D. *Gorgon: Paleontology, Obsession, and the Greatest Catastrophe in Earth's History.* New York: Viking Press, 2004.

Wignall, P. B., and R. J. Twitchett. "Oceanic Anoxia and the End Permian Mass Extinction." *Science* 272 (1996): 1155–58.

CHAPTER 7: TRIASSIC SPRING

Bolton, Barry, and Ian Gauld. *The Hymenoptera*. New York: Oxford University Press, 1988.

Jones, T. D., J. A. Ruben, L. D. Martin, E. N. Kurochkin, A. Feduccia, P. F. A. Maderson, W. J. Hillenius, N. R. Geist, and V. Alifanov. "Non-Avian Feathers in a Late Triassic Archosaur." *Science* 288 (2000): 2202–5.

Key, K. H. L. "Phasmatodea (Stick-Insects)." In *The Insects of Australia*, vol. 1, edited by Ian D. Naumann, 394–404. Carlton: Melbourne University Press, 1991.

Labandiera, Conrad C., and J. J. Sepkoski Jr. "Insect Diversity in the Fossil Record." *Science* 261 (1993): 310–15.

Ross, Edward S. "Embioptera, Embiidina (Embiids, Web-Spinners, Foot-Spinners)." In *The Insects of Australia*, vol. 1, edited by Ian D. Naumann, 405–9 Carlton: Melbourne University Press, 1991.

Sereno, P. C. "The Evolution of Dinosaurs." *Science* 284 (1999): 2137–47.

———. "The Origin and Evolution of Dinosaurs." *Annual Review of Earth and Planetary Sciences* 25 (1997): 435–89.

CHAPTER 8: PICNICKING IN JURASSIC PARK

Chiappe, L. "The First 85 Million Years of Avian Evolution." *Nature* 378 (1995): 349–55.

Dingus, L., and T. Rowe. *The Mistaken Extinction: Dinosaur Extinction and the Origin of Birds*. New York: W. H. Freeman, 1998.

Gauthier, J., and L. F. Gall, eds. *New Perspectives on the Origin and Evolution of Birds*. New Haven, CT: Yale University Press, 2001.

Holland, William Jacob. "The Skull of *Diplodocus*." *Memoirs of the Carnegie Museum* 9 (1924): 379–403.

Ostrom, J. H. "Bird Flight: How Did It Begin?" *American Scientist* 67 (1979): 46–56.

Padian, K., and L. M. Chiappe. "The Origin of Birds and Their Flight." *Scientific American* 278 (1998): 38–47.

Quicke, Donald L. J. *Parasitic Wasps*. London: Chapman and Hall, 1997.

Vilhelmsen, Lars, and Giuseppe Fabrizio Turrisi. "Per arboretum ad astra: morphological adaptations to exploiting the woody habitat in the early evolution of Hymenoptera," Arthropod Structure and Development 40 (2011): 2–20.

CHAPTER 9: CRETACEOUS BLOOM AND DOOM

Bandi, C., M. Sironi, G. Damiani, L. Magrassi, C. A. Nalepa, U. Laudani, and L. Sacchi. "The Establishment of Intracellular Symbiosis in an Ancestor of

Cockroaches and Termites." *Proceedings of the Royal Society of London Series B* 259 (1995): 293–99.

Desalle, R., J. Gatesy, W. Wheeler, and D. Grimaldi. "DNA Sequences from a Fossil Termite in Oligo-Miocene Amber and Their Phylogenetic Implications." *Science* 257 (1992): 1933–36.

Doyle, James A. "Molecular and Fossil Evidence on the Origin of Angiosperms." *Annual Review of Earth and Planetary Sciences* 40 (2012): 301–26.

Grandcolas, P., and P. Deleporte. "The Origin of Protistan Symbionts in Termites and Cockroaches: A Phylogenetic Perspective." *Cladistics* 12 (1996): 93–98.

Grimaldi, David. "A Fossil Mantis (Insecta: Mantodea) in Cretaceous Amber of New Jersey, with Comments on the Early History of the Dictyoptera." *American Museum Novitates* 3024 (1997): 1–11.

Lo, N., G. Tokuda, H. Watanabe, H. Rose, M. Slaytor, K. Maekawa, C. Bandi, and H. Noda. "Evidence from Multiple Gene Sequences Indicates That Termites Evolved from Wood-Feeding Cockroaches." *Current Biology* 10 (2000): 801–4.

Raup, David M. *The Nemesis Affair: A Story of the Death of Dinosaurs and the Ways of Science.* New York: W. W. Norton, 1986.

Schmidt, Justin O. "Hymenoptera Venoms: Striving Toward the Ultimate Defense Against Vertebrates." In *Insect Defenses: Adaptive Mechanisms and Strategies of Prey and Predators,* edited by David L. Evans and Justin O. Schmidt, 387–419. Albany: State University of New York Press, 1990.

Thorne, Barbara L. "A Case for Ancestral Transfer of Symbionts between Cockroaches and Termites." *Proceedings of the Royal Society of London Series B* 241 (1990): 37–41.

———. "Evolution of Eusociality in Termites." *Annual Review of Ecology and Systematics* 28 (1997): 27–54.

Thorne, Barbara L., and James M. Carpenter. "Phylogeny of the Dictyoptera." *Systematic Entomology* 17 (1992): 253–68.

Thorne, Barbara L., David A. Grimaldi, and K. Krishna. "Early Fossil History of Termites." In *Termites: Evolution, Sociality, Symbioses, Ecology,* edited by T. Abe, D. E. Bignell, and M. Higashi, 77–93. Dordrecht: Kluwer, 2000.

Thorne, Barbara L., and James F. A. Traniello. "Comparative Social Biology of Basal Taxa of Ants and Termites." *Annual Review of Entomology* 48 (2003): 283–306.

CHAPTER 10: CENOZOIC REFLECTIONS

Erwin, Terry L. "How Many Species Are There? Revisited." *Conservation Biology* 5 (1991): 330–33.

———. "The Tropical Forest Canopy: The Heart of Biotic Diversity." In *Biodiversity,* edited by Edward O. Wilson, 123–29. Washington, D.C.: National Academy Press, 1988.

———. "Tropical Forests: Their Richness in Coleoptera and Other Arthropod Species." *Coleopterist's Bulletin* 36 (1982): 74–75.

Hutchinson, E. E. "Homage to Santa Rosalia or Why Are There So Many Kinds of Animals?" *American Naturalist* 93 (1959): 145–159.

Johanson, Donald C., and Maitland A. Edey. *Lucy: The Beginnings of Humankind.* New York: Simon and Schuster, 1981.

Labandeira, Conrad C. "The Fossil Record of Insect Extinction: New Approaches and Future Directions." *American Entomologist* 51 (2005): 14–29.

Lovejoy, C. O. "Reexamining Human Origins in Light of *Ardipithecus ramidus.*" *Science* 326 (2009): 74–78.

Shaw, Scott R. "Essay on the Evolution of Adult-Parasitism in the Subfamily Euphorinae (Hymenoptera: Braconidae)." *Proceedings of the Russian Entomological Society, St. Petersburg* 75 (2004): 1–15.

White, T. D., B. Asfaw, Y. Beyene, Y. Haile-Selassie, C. Lovejoy, G. Suwa, and G. WoldeGabriel. "*Ardipithecus ramidus* and the Paleobiology of Early Hominids." *Science* 326 (2009): 64–86.

POSTSCRIPT: THE BUGGY UNIVERSE HYPOTHESIS

Armitage, A. "The Cosmology of Giordano Bruno." *Annals of Science* 6 (1948): 24–31.

Gould, Stephen Jay. "The Evolution of Life on the Earth." *Scientific American* 271 (1994): 85–91.

Greene, Brian. *The Elegant Universe: Superstrings, Hidden Dimensions, and the Quest for the Ultimate Theory.* New York: Vintage Books, 1999.

Orgel, Leslie E. "The Origin of Life on the Earth." *Scientific American* 271 (1994): 77–83.

Rebek, Julius Jr. "Synthetic Self-Replicating Molecules." *Scientific American* 271 (1994): 48–55.

Sagan, Carl. "The Search for Extraterrestrial Life." *Scientific American* 271 (1994): 93–99.

Silk, Joseph. *A Short History of the Universe.* New York: Scientific American Library, 1997.

Taylor, G. Jeffrey. 1994. "The Scientific Legacy of Apollo." *Scientific American* 271 (1994): 40–47.

Taylor, S. R. "The Origin of the Moon." *American Scientist* 75 (1987): 469–77.

Yates, F. *Giordano Bruno and the Hermetic Tradition.* Chicago: University of Chicago Press, 1964.

Index